# The Step 1 Method

## A Step by Step Guide to Success on the USMLE Step 1 Exam

# The Step 1 Method

## A Step by Step Guide to Success on the USMLE Step 1 Exam

Dan Gebremedhin MD, MBA

ISBN 978-1-300-35687-5

*This book is dedicated to the students and faculty who have supported the program through the years. Your feedback and encouragement is what keeps us going.*

"Carefully watch your thoughts, for they become your words. Manage and watch your words, for they will become your actions. Consider and judge your actions, for they have become your habits. Acknowledge and watch your habits, for they shall become your values. Understand and embrace your values, for they become your destiny."

- Mahatma Gandhi

# Contents

# Section 1

# Overview and Content Strategy

# Chapter 1

## An Overview of the USMLE Step 1 Exam

### What Is the USMLE Step 1 Exam?

For many reasons, the United States Medical Licensing Exam (USMLE) Step 1 Exam is the most challenging of all three USMLE Step exams. From the expansive basic science breadth of the exam, to the weight placed on it by residency program directors, Step 1 is a major milestone on the path to practicing medicine in the United States. Students' responses to this exam vary widely: Some will be paralyzed by fear; some will be galvanized by the challenge; and some will choose *La Belle Indifference* approach, minimizing the importance of the test. It is this last group I worry about the most because I believe this approach is a defense mechanism used to cope with the weight of the exam. After taking the exam, these students usually regret their indifferent approach to it. They may not pass, or they may barely pass, forced to live with a score that is well below their potential.

Consequently, students taking the Step 1 Exam are mired in a multitude of conflicting emotions. To combat this, we take a minimalist approach to setting goals. We believe that every student should attempt to reach their potential on the exam. Some students have the potential to score above 250, while others will work hard to pass the exam comfortably on the first attempt. Both goals are admirable. We have tried to create a program that enables students to reach their potential on the exam, whatever it may be.

There is much debate in the educational community regarding learning styles. Some students prefer to learn in a particular way and some instructors and programs prefer to teach in a particular way. Educators have done research to find the "most effective" way to learn information. If you survey our program, you will see overlap with some leading educational methodologies, but we didn't rely only on that research to come up with the *Step 1 Method*. We created our program from first-hand experience, beginning in 2004, year after year, working with thousands of students. Our program today looks very different from the way it did eight years ago. Annually, we objectively measure our results and obtain feedback from participants to determine what worked and what didn't. In 2012, we contracted with four US medical schools and helped a cohort of 600 students prepare for the exam. These results are becoming increasingly interesting as we are able to measure what we have always known: it is possible to prepare for the USMLE Step 1 Exam. We are showing that proper preparation for this exam correlates at a much higher rate than a top MCAT score.

We've seen the resistance to changing one's learning style for a particular exam. In spite of past, repeated, stellar results, every student believes their situation is different. I don't argue with students when they say they "prefer" to study with flashcards or watch online lectures. Students prefer to remain in comfortable scenarios. They want to do "what they've always done" because it has gotten them this far. Our program challenges you to enter new and uncomfortable territory. We want you to think about topics in a way that you may never have. Coincidentally, we teach you to think about material the way the Boards' examiners do. We believe this is the most efficient way to answer questions correctly on this test.

# The Material

The USMLE Step 1 Exam (or the "Boards") covers the basic sciences that are the subject of the first two years of US-based medical schools. This is not to say that all US medical schools will teach to every topic on the USMLE Step 1 Exam. On the contrary, the organization that creates the Boards, the National Board of Medical Examiners (NBME), develops the exam independently of

US medical schools. It annually publishes an informative document of more than 50 pages that lists the 550+ individual topics the test may cover. US medical schools, on the other hand, create their curriculums independently, with one eye on the Boards, but with another eye on academic integrity. Many US medical schools go out of their way to ensure that they "don't teach to the test."

Realistically, if you went to a US medical school, you will have covered a majority of the topics that are tested on the Boards. A faulty assumption is that if you cover a topic in med school, you will have learned it to the depth required by the NBME and as it is reflected in the questions on the Boards.

I feel less strongly about making the same assertion for international medical schools as curriculums vary across the world. But, if you studied at an international school that focuses on "Western medicine" and trains physicians who frequently come to practice in the United States, you are probably safe in assuming that you also have been exposed to a majority of the topics on the USMLE Step 1.

The purpose of this text is not to list all the high-yield topics on the Boards. There are various texts that do an adequate job of that, and we mention a few of them that are known as "central texts" and discuss them later in the book. This text is meant to serve as a guidepost in the treacherous and high-cost waters that is studying for the Boards.

## The Significance

There is much debate on the real significance of the USMLE Step 1 Exam. Having served on an Internal Medicine residency selection committee, I can tell you the result on this exam is not *the most* important criteria when evaluating a candidate. There are much more important criteria to consider when determining who will make great physicians and/or medical researchers. Namely, these include clinical grades in the third and fourth years and letters of recommendation from your dean regarding your character and academic integrity. If these main criteria are in ill repute, then a great Board score won't likely resurrect a sorely damaged reputation.

# Why Won't a Score of 260 Cure All Ails?

The nature of the selection process in most US residency programs is that there are fewer spots than there are applicants. Residency selection committees use their preferred method of judging the future potential of candidate physicians in a matter of one to two months, which is very hard to do. Committees can make mistakes when judging the degree of "fit" between the candidate and the culture of their program. If such a miscalculation is made, then the program will have to deal with this incongruence for the next three to five years of residency. Culture is extremely important in a residency program because attending physicians are reliant on residents to uphold the integrity of their clinical practice. A resident who is seen as untrustworthy—for whatever reason—can be very detrimental to the entire residency program. For these reasons, selection committees are wary to take "risks" on applicants.

There are always very qualified applicants who do not get into their top choice of residency. Naturally, there are varying levels of the "competitiveness" of residency applications. Thus, the residency that is your second choice is likely someone's first choice. And if you got a 260, but your other marks are highly suspect, you are seen as "risky." A student with a 235–245—an above-average and good score—with outstanding clinical marks and recommendations is seen as a "safe bet." Although this is a gross simplification of the process, it is these sorts of value judgments that selection committees must make. This is also a good example of the limited role that the Board exam score can play.

The USMLE Step 1 Exam probably plays the largest role in very competitive residencies. "Competitive" is defined by the ratio of the number of available residency spots to the number of candidates applying for those spots. Historically, most students consider specialties such as dermatology, radiology, ENT, etc. as the most competitive, but even the top internal medicine programs can also be considered "competitive" based on the above definition. Further, many students have a desired geographical region of residency, which can also make that residency "competitive." Because these residency directors are overwhelmed with applicants for a limited number of spots, they need a screening mechanism with which they can limit the field. Most often, the USMLE Step 1 score serves as that screening mechanism.

The USMLE Step 1 Exam is objective. The same set and style of questions are administered to all students equally. The problem with medical school grades is that there is a wide disarray of grading scales. Some schools are Pass/Fail, some use a modified tiered grading system, and some remain on the A, B, C scale. Some schools rank their students, while other schools do not. In this world of variability, the USMLE Step 1 Exam serves as a level playing field and is readily used by program directors for this purpose.

To be clear, I do not believe the USMLE Step 1 Exam has the ability to determine who will be a great physician. On the contrary, I believe the USMLE Step 1 Exam measures the ability to answer questions about basic medical science that are asked in a particular style. Further, I believe one's score on the USMLE Step 1 Exam has a direct correlation to two things: knowledge of the basic sciences and proper preparation. I have coached many students with MCATs in the 20s who scored in the 250s on the USMLE Step 1 Exam, but I have also seen students with MCATs in the high 30s and 40s—who also had top basic science grades—score well below their potential on the USMLE Step 1 Exam because they lacked proper preparation.

In terms of setting goals, I encourage all students to work hard and reach their potential on this exam. A student's potential is based on many things: namely, their knowledge of the basic sciences, their score goal and motivation, their test-taking abilities, and the time allotted to preparation. Given that many students don't know what specialty they want to go into, I encourage them to do as well as they can to leave their options open. In our Preparation Strategy section, we cover how to determine what your score potential is and how to gauge your score trajectory based on our past research.

## The Timed Exam

The USMLE Step 1 Exam is a computerized, timed test made up of seven blocks of 46 questions. You have 60 minutes to finish each block of questions; that is on average 1 minute and 18 seconds per question. There is no penalty for guessing. You have to complete one 60-minute block at a time and cannot browse among different blocks of questions. Over the course of the eight-hour day, you have the ability to take a total of 45 minutes of break time in between blocks. You can choose to take this time at your leisure.

If this all sounds very intimidating, it should. To the untrained student, this exam can be a catastrophe. Many times, unprepared students will run out of time at the end of every block, leaving several questions unanswered, which is demoralizing and which leaves you dejected and defocused as you enter the next block of questions. As you can see, doing poorly on the exam is often a self-propagating occurrence.

Fortunately, as you will see in our Preparation Strategy section, we have developed strategies that will enable every student to finish the exam comfortably and on time. Even students with the highest amounts of test-taking anxiety and those who have problems with standardized tests can train themselves so that they will have time at the end of every block to review their answers.

# The Grading

The way that the NBME grades the USMLE Step 1 Exam is somewhat of a black box. There is no publicly available, official proclamation of how it grades the exam. I've read different publications and documents produced by the NBME that have allowed me to piece together a best guess of how the test is graded. In short, the NBME compares students on an individual question basis to determine a question's difficulty level. For example, if 75 percent of students get a particular question wrong, then that is a difficult question. The NBME will then compile an exam with an even distribution of questions that fall all along the difficulty spectrum. This enables the NBME to very easily use an absolute scale of the percent correct to a corresponding three-digit score (e.g. 240). I would best describe this as a modified curve. The passing score has the ability to change based on the performance of that year's students. In 2012, the passing score was 188. This passing score has remained at this level for the past three to four years. The average score, on the other hand, has steadily risen for the past eight years. In 2004, the average score was in the 215 range. In 2012, the average score rose to 224.

Most commonly, the three-digit score is used to represent a student's score. Historically, an additional two-digit score on the 100 scale was also used to describe the student's score in a language that programs could understand. For example, the passing score was always pegged to the two-digit score of 75, while higher scores of 245 and up were given the two-digit score of 99. As of 2012, the NBME has decided to discontinue the two-digit score reporting.

# Registering for the Exam

To register for the exam, go to http://www.usmle.org/apply/. In 2013, the exam will cost $560. You must designate a three-month period in which you wish to take your exam. Once you are approved for that three-month period, you can call the Prometric testing center or schedule online a specific date within those three months, depending upon availability at specific locations. The fee for extending your three-month testing period by another three months is $65. An additional extension will require re-application and may require resubmission of total fees. It is best to gain familiarity with the NBME/USMLE and Prometric test center websites as you may wish to change your exact test date within your eligibility period.

# The Time Allotted for Preparation

US medical schools differ in the dedicated time they provide to their students after the second year to spend studying for the Boards. Most schools give their students between four and eight weeks to study for the Boards. In the extreme case, some schools give their students only three weeks to study for the exam, but they may have a light last quarter/semester of classes. Some schools allow their students to take the exam at some point between the beginning of their second year and the end of their third year.

The *Step 1 Method* is designed to help you prepare for this situation of open-endedness. We have definitive ideas about how long you should study for this exam. Many students discover our program only months before the exam and fret that we actually recommend that students start studying six months out. Although we recommend starting early in a regimented fashion, we know that the world is not perfect. It is never too late to start studying the right way. If you learn about our program ahead of time, please do yourself a favor and don't rush your preparation for this exam. Your score potential will correlate directly with how long you give yourself to learn the strategies and deepen your understanding of the basic medical sciences. It takes time to transition from memorization to the deep understanding that the NBME requires.

In our Study and Preparation Strategy sections, we cover when you should start working on questions and at what rate you should do them. We cover how long your Intensive Study Period should be. Of course, your Step 1 study exists within the confines of your life. You have to make decisions based on your situation and what time you are able and willing to invest. Hopefully, we'll convince you that this protected time will never come again, and you should use it to the best of your abilities to become the best doctor that you can be.

# The Resources

In our Study Strategy section, we outline which resources you should use to prepare for this exam. Every student has different weaknesses and strengths, therefore their resources should be tailored to their needs. That said, we believe that all students should begin with a core of a few resources and then cautiously broaden their use of supplementary resources to address their special needs. In short, we recommend that all students obtain a six-month subscription to a leading Qbank (as of 2012, USMLE World is the best on the market); a central, comprehensive text (e.g. *First Aid*); and two to three supplementary subject/system-based texts to address a student's main weaknesses. Our Content and Study Strategy sections address how you should use these resources optimally to get the most out of them. In our Study Strategy section, we provide a methodology for how to think about other resources such as online lecture and audio courses, flash card and slide banks, and other supplementary texts.

The fundamental resource that most students overlook is the practice test. In our Preparation Strategy section, we cover which practice tests to use and when and how to use them. In most cases, students should use only the most recent NBME Basic Sciences Self-Assessment forms (11, 12, 13) with enhanced feedback. These are the only practice tests that have been statistically proven to provide diagnostic correlation with USMLE Step 1 Exam performance.

# Self-Preparation or Review Course

The *Step 1 Method* was designed as a way for students to teach themselves the material. When we perform engagements with our client schools, we simply guide the student body through the *Step 1 Method* online course. In our Study Strategy section, we cover an in-depth analysis of why we believe the *Step 1 Method* is successful. The fundamental assumption of our program is that medical students are smart enough after having gone through two years of medical school to review the information they've covered and learn new topics that they encounter. We believe that active learning through the process of working questions and targeted reading leads to a higher level of retention and a deeper level of understanding. Sitting through lectures, on the other hand, is a passive form of learning, and there is no guarantee that the lecture material will go into a deep enough level to answer the three- and four-step questions the NBME asks. Further, when students sit in a lecture, whether live or online, they are not in control. They learn what the instructor teaches. The Intensive Study Period should be an exploration of each student's individual weaknesses. Without the ability to control the flow of information, students usually do not go deep enough into their areas of weakness.

There are a number of organizations that will invite a parade of instructors for a three-week course and charge students more than $4,000 for entry into such "intensive prep" programs. These courses take up a predominant portion of the student's day and energy. We believe students should be spending a majority of their time working questions the right way. If students are spending a majority of their time in lectures, they will never reach their score potential. For most motivated US medical students and IMGs, I believe these courses are not the right prescription. For IMGs whose curriculum was drastically different from the US curriculum, it may be beneficial. Many times, students remark that they don't have the self-control to study several hours a day and look to these courses for the external control. To those students, I have little to say. I would hope that these students would have the discipline to be able to study on their own for at least eight hours a day. There are many more creative, effective, and less expensive ways to create external control, for example, studying in conjunction with a group or studying in an official/austere location such as a library.

I will agree that our program is only for motivated students. If you are not motivated to reach your score potential and not willing to work hard, you need not sign up for our program. Granted, you will need to be your own drill sergeant, but you will have detailed instructions on what you have to do. With our increasingly available online tools, you will not feel as though you are doing it alone. Furthermore, with our proven methods you can eliminate fear from the equation.

The vast majority of US medical students study on their own, but we are seeing an increase in the popularity of content-based online lecture courses. This is likely because of the comfort and familiarity that comes from watching video lectures. I agree that watching and understanding a lecture is much easier than getting multiple questions wrong and learning from your mistakes while you tackle your weak areas. We see the same problem with online lecture courses as we do with the live lecture courses. You must look at how you spend a majority of your time. If you spend a majority of your time attempting to passively learn in areas that are not your weakest, then you can hardly expect to get a great score.

I must point out the difference between our online strategy-based lectures and content-based online lectures. Our online lecture library focuses on teaching you the strategies to study and answer questions more effectively. We do not try to teach you the information via the videos; we teach you how to teach yourself the material at the level of understanding required by the NBME. We've also found the best way to teach strategy is through simulation and not just text.

# Chapter 2

## The *Step 1 Method*©

### The Story behind the *Step 1 Method*

They often say that necessity is the mother of invention. Created in 2005, "The Method" filled the void for students who wanted a proven framework for how to study for the USMLE Step 1 Exam. Normally, students have to depend on anecdotal advice from upperclassmen or anonymous posts on chat forums regarding how to prepare for the exam. There are several problems with this common scenario: There is no way to confirm that the methods used by other individuals actually worked because scores are rarely shared among students. Usually, there is a rumor that a particular upperclassman "did really well, and they said, 'do xyz.'" Further, even if a student did well, it is impossible to determine if the resource used was responsible for a score increase.

Students find themselves using review guides like BRS and *Rapid Review* for individual subjects throughout their first two years. They assume that the way to a great score is to combine reading these supplementary sources for each subject along with "memorizing *First Aid*." Students also incorporate the use of a question bank when they prepare for the exam, focusing on finishing questions as fast as possible and often doing them in a haphazard fashion. They work the questions in "random" subject mode and time themselves, trying to increase their speed. Then, they get depressed when their percent correct is lower than they would like. Because of this negative experience, they will put off working more questions and go back to reading so that they can "learn" the information prior to working questions. This cycle continues. If students start early, they hope to finish one Qbank so that they can move to a second Qbank and fly through those questions as well. They have heard that the more questions the better.

Some students may feel more secure if they use flash cards so that they can memorize the many "facts and details" the Boards focuses on. Other students cling to online lectures that they paid $1,000 for earlier in the year after they heard from other friends who purchased the series and really liked it. Because they purchased the series, they feel the need to use it.

As you can imagine, students using multiple sources at different times are pulled in all directions. They get overwhelmed with all the information they encounter. Their natural response is to go faster and study longer so that they can get through it all. Their time spent studying for the Boards is a blur and a mad dash. The common result is that students spend months studying for this exam and are dumbfounded when their score is far lower than they expected.

This was the scenario back in 2004, and it's still the story for tens of thousands of students every year. In 2010, 10 percent (2,300) of all US test-takers failed this exam. In that same year, 39 percent (7,380) of IMGs failed the exam.[1] These are just the numbers for students who failed. I know that there are many other students who passed but scored well below their potential. Having spent the last seven years working with students, I know it's not due to a lack of effort; it's usually the lack of proper preparation.

# The Beginning

I worked as an MCAT teacher for the Princeton Review in college and med school. The Princeton Review was a very "strategy–based" upstart shop when compared to the content-heavy Kaplan. Those years taught me how to instruct students how to prepare for the strategy of the exam and not just the content. When thinking about how to best study for the Step 1 Exam, I initially went about it as I previously described. I spoke to the upperclassmen who were rumored to have done well. Whatever they remembered from the Board study period a year ago, they gladly shared. Some classmates

---

[1]    Source USMLE.org

introduced me to the leading medical student forums. I remember feeling like a new world was opened to me: all these people were posting advice about the best resources to use as well as their scores to prove that they were trustworthy. I even participated in a few discussions, but I quickly became disenchanted with the environment because of the arguments that dominated the site. I found myself with too many resources, too much advice, and not enough time.

The one thing I knew from my Princeton Review days was that simulation was the key to success. If you could become extremely comfortable with the test, you would likely do better. So, I spent a majority of my time working questions. I started a virtual message board within my medical school class so that people could share what resources they were using or questions they had. We created a community environment that fostered collaboration.

Eventually, I settled on what I thought were some best practices:

1.  Get through the question bank at least once, and begin to go through it a second time.

2.  Use your *First Aid* along with the Qbank to write down new ideas and clarification learned in questions.

3.  Use subject-based review books in your areas of weakness.

4.  Take NBME practice tests to track your progress.

I shared this set of simple best practices with our class. We ended up with a 98 percent pass rate and an average score of 215, both marks at or above the national average for the first time in school history. I scored a 256 and was happy with my score. I turned the simple best practices into a seminar and coached the following classes on how to do well on the exam. The second year of the program, our school finally achieved the elusive 100 percent pass rate and had a class average higher than that of the national average. The significance of this statistic was that our school had MCAT scores that were below the national average. It has been oft cited that a student's Step 1 score correlates directly with their MCAT score. These results showed that the intervention of these best practices enabled the students to perform much better than expected on the Step 1 Exam.

Soon after these results became public, I began to get invited to medical school conferences and individual medical schools to speak about the best practices, which became known as the *Step 1 Method*.

A few years later, a medical school in the northeast United States officially contracted our services to prepare their entire medical school class for the exam. They immediately improved their pass rate and average score. Since then, we have branched out to four US medical schools and are still growing and learning. Our program has grown out significantly from those initial best practices to address the problems our students find most pressing.

As you will see, the best practices that we started with in 2005 are now commonplace among student study plans, and the average score has risen nine points. In response, we have had to be innovative with the *Step 1 Method 2.0* in our Content and USMLE Strategy sections. We've also continued to refine our Preparation and Study Strategy sections. We are focusing more on data collection and objectively measuring the effect of our program. We are utilizing the best resource we have—a cohort of hundreds, and soon thousands, of students who are taking the exam annually. Our goal is simple: we try to get better every year to give our students the best possible USMLE experience and the best chance to reach their score potential, whatever that is.

# An Overview of the Program

The *Step 1 Method* is divided into four major strategies that encompass a student's entire preparation program. We will briefly outline the sections here.

## Content Strategy

This section covers the vital question: "What do students need to know for the Boards?" We answer this question by focusing on how the NBME thinks about information on this test. Using our *Step 1 Method* Framework, we show students how to think about all information on this test. This framework focuses on categorization, integration, and retention of the material.

## USMLE Strategy

A major focus of our program is on showing our students how to think like the NBME. This section teaches students about the style of the USMLE questions as well as how NBME creates its questions. By providing insight into the question-creation process, students can then anticipate and prepare for potential questions. This section also provides students with tangible strategies to increase their understanding and their speed of answering the challenging USMLE questions. With these strategies, students report increased confidence and decreased anxiety during the exam. Our goal is for our students to have total familiarity with the exam. One dictum heard often at *Step 1 Method* rallies is that on the Boards, "You want to have to figure out as little as possible, and recognize as much as possible."

This section also focuses on developing test-taking strategies for the USMLE. We tackle vital issues such as timing on the exam, canceling out wrong answer choices, and how to combat fear of standardized tests.

## Study Strategy

This section answers the question: How can one learn the content for the Step 1 Exam in a manner that focuses on high retention and reproducibility of the information? Further, which resources should be used to reach this goal?

We focus on practical advice and guidance to show how to strategically prepare for the exam throughout the second year and during the intensive study period. By providing timeline guidance, students can focus and excel in the second-year curriculum without worrying about planning their Step 1 prep. We walk you through our active learning format known as the *Step 1 Method* Format of Doing Questions. We provide guidance on the resource-selection process. We show you how to plan your Intensive Study Period and give you access to our Study Schedule Wizard software that does all the work for you.

## Evaluation and Motivation Strategy

This section focuses on the weeks leading up to the exam. It helps you answer the questions: How do you accurately assess performance? How do you use this assessment constructively to attack weaknesses? How do you create an effective study plan after learning this information?

We show you how to use the NBME practice tests to track your performance and attack your weaknesses to reach your score potential. You will know exactly what to expect from the Boards well before you walk into the exam.

This section also deals with the emotional side of preparing for the Boards. We show you how to effectively set appropriate goals, plan schedules, and cope with the day-to-day stress this test brings. We also highlight the importance of focus during both prep and exam time. Through mental visualization exercises, students can naturally decrease the amount of anxiety brought on by the USMLE Step 1 Exam.

# Chapter 3

## Getting into the Right Mindset

### Introspection

"What one thing could you do (that you aren't already doing now) on a regular basis, that would make a tremendous positive difference in your personal and professional life?"

Taken from *The 7 Habits of Highly Effective People* by Stephen R. Covey

We created the *Step 1 Method* to be more than just a test-prep tool. We wanted to give students the skill sets to succeed on more than just the USMLE Step 1 Exam. If you apply the principles highlighted in the program, you can assure yourself success on USMLE Steps 2 and 3, your specialty Board exams, and any other standardized tests you take in the future. If taken to heart, the program will reshape the way you look at your time; it will force you to reexamine your goals and your priorities. Limitless potential lies within you. You have achieved great things up to this point, but in order to get to the next level, you must ask tough questions and examine your daily pattern of habits.

Have you ever found yourself going into a test worried, feeling stressed, wishing you had more time to prepare? If so, you've fallen into a common trap with the majority of other students. Talented individuals tend to get by on the minimum amount of work necessary to succeed, but not to excel. In highly competitive environments such as the USMLE Step 1 Exam, if you never give it your all, you will never get to where you want to go.

You should never have to accept where you end up in life; you should always have the power to choose.

Today, you will be given the information necessary to succeed on the Step 1 Exam—the choice is then yours. Will you take the next step and reach your potential on the USMLE Step 1 Exam?

# Keys to Success

Before we get into the contextual basis of Step 1 test preparation, we must first deal with mental preparation, which will undoubtedly be the toughest and most significant test of your medical career. From here on out, most tests will be pass or fail; if scored, the score will bear minimal importance. The USMLE Step 1 Boards, and more importantly, the preparation for the exam, are both mentally daunting. You must be prepared mentally before you begin such an arduous journey.

# Desire

In my experience, many students turn off when thinking about the Step 1 Exam and specific score goals. For some, a score is not an inspiring or motivating goal. Thus, while it is fine to choose a score goal and chase it, we'd like you to focus on a different goal. You've spent the last two years learning the most basic medical science you'll ever know. From now on, a majority of your education will be clinically based. You will never get another opportunity to have protected time to review and learn basic science. This is surprising because every field of medicine, from surgery to radiology, involves basic medical science. Those who have a mastery of both clinical and basic sciences will excel in their careers and take better care of their patients. So, when thinking about why you are spending 10-hour days studying for this test, focus on being the best doctor you can be. The knowledge you gain and consolidate in the upcoming months will aid you in your clinical rotations and will cure and save the lives of your future patients.

# Plan and Action

The key to accomplishing anything is to create plans for its achievement and then to put those plans into action. It is great to have a goal, but if you don't have explicit directions for the completion of that goal, you will invariably waffle along— sometimes making progress, and other times not. One mistake many students make is collecting a multitude of resources to use for their Step 1 study. When they hear peers say a resource is good, they feel pressured to purchase it and use it. Often, these students are pulled in so many directions and go from audio lecture to flash cards to reading to questions without a clear plan. They certainly stay busy, but they never gain the feeling of accomplishment or mastery of any one topic. Because they are working hard and staying busy, they feel confident prior to taking a practice test but are shocked when the result is well below what they expected.

- In the *Step 1 Method*, we believe that you cannot be successful unless you have a plan and execute that plan. Ironically, having a set plan is most important in the event that you are not performing at your desired level; the only way to improve is to diagnose where you have gone wrong. This is impossible if you have no idea where you have been.

# Evaluating Your Performance

Once you create plans and start putting them into practice, you must determine whether your plan is working. You must be constantly reevaluating yourself to determine if you are making progress. When thinking about reaching your score potential, you must constantly reevaluate your weaknesses and where they stand. It does not make sense to spend significant time on your strengths because they can get only incrementally better. By attacking your weaknesses repeatedly, you have the opportunity to make great strides and huge score improvements. Conversely, if it has been some time since you covered a strength, it can also become a weakness. Thus, the assessment and review process must be iterative.

One all-too-common mistake students make is not using practice tests the right way. Students either take the wrong tests (those made by the Qbanks), avoid taking them, or when they do take the right tests (NBME CBSSAs), they don't use them the right way.

- In the *Step 1 Method*, you will spend a large portion of your time in the Intensive Study Period assessing your weaknesses through scheduled practice tests and learning from your mistakes in an iterative fashion.

## Refine Your Approach

If you want to reach your score potential, you must learn from your mistakes. Learning from your mistakes requires self-inventory and awareness. We find that students work hard but do not take the time to notice when they're making mistakes. A great example of this is when a student takes an NBME practice test, receives a score, and surprisingly chooses not to purchase "enhanced feedback," which is an additional report where the NBME will note which actual questions you missed. Even more surprising, we encounter many students who will purchase the enhanced feedback but do not look at the report. The main excuse these students have is that they are "too busy" and have too much work to do. We also commonly see students speed through questions and gloss over major weaknesses without stopping to hammer out the *Step 1 Method* Framework for these weaknesses.

- In the *Step 1 Method*, we expect you to make mistakes, but you must learn from them. You must take the time to acknowledge where you are weak and hammer out those weaknesses before they show up on the test.

## Define Your Goal

When attempting to accomplish anything, it is essential to have a specific object, task, or accomplishment on which to focus. If you remain vague in defining your goals, you will never know when you achieve them. More importantly, if you don't have something specific on which to focus, it will be difficult to stay motivated. It is easy to say, "I want to do well." What is "well"? Where does the "okay" end and "well" begin? Without strict guidelines to wrap your psyche around, you will never develop the necessary motivational energy to withstand four to five weeks of intense study. You will quickly get fed up and tired of the same routine unless you have a strong and foreseeable endpoint or finish line. Your goal is the light at the end of the tunnel, and you must focus on it throughout the study period or you will find yourself confused and lost when there is no time to waste for uncertain, mixed-up feelings.

Some schools only allow their students a few weeks to prepare for the exam; a few weeks won't produce optimal results for most students. Therefore, you must go above and beyond the bare minimum effort that your school expects of you. This requires you to be proactive and determined to see results on your own accord. The only way to reach the next level of performance is to step up your effort to the next level. You have to work harder than you ever have before. I know this may seem daunting, but trust me and anyone else who was truly satisfied with their score—the hard work is definitely worth it. It will make you a better student, a better physician, and overall a better person. Later in the method, we recommend a study schedule that requires starting early and a minimum of four weeks of intense preparation, assuming your school gives you that much time.

The only way to elicit maximum effort from oneself is to set a worthy goal. You have to imagine your ideal score; a score that, if you received it, would make you say "Wow!" Furthermore, we previously mentioned the technique of mental imagery: Imagine how hard someone would have to work to get that score, and work that hard. Imagine how diligent, thoughtful, and proactive a person receiving that score would be. Become that person today. How would that person carry themselves? The *Step 1 Method* was initially created as the collection of best practices of people who have achieved amazing results. Become an eternally curious student, always looking for the best way to accomplish tasks related to your medical education.

If your dream score is truly high, there is a good chance that you may not meet it. But if you work as hard as you intend to, your actual score will be very close to that dream score, and therefore an excellent score indeed. Often, people aspire for a vague goal: to do "good" or "well," or worse, "to pass." That is the wrong the idea. If you shoot for mediocre and vague goals, you will receive mediocre results. But as the saying goes, if you shoot for the moon, you'll still end up in the stars.

The USMLE Step 1 Exam is the single-most important exam in your medical career. No other exam will open or close doors or make or break your dreams of having a medical career in a particular profession or a particular place. Every year, the national average score increases slightly. In the past, a 230 was seen as competitive. Now, that universally "competitive" score is considered to be a 240. Needless to say, the higher the score, the more competitive the applicant will be. Remember, USMLE Step 1 scores are used as a cutoff mechanism for residency applications, meaning that many programs will not interview individuals with USMLE Step 1 scores lower than a particular "cutoff." To be clear, the USMLE Step 1 score alone will not gain acceptance for a particular applicant, but it will open the door for the applicant to share their personality and other accomplishments with the program via interview.

More accurate matching data is now available at the NRMP website in "Charting Outcomes in the Match." This details average Step 1 scores, time spent in research, and other characteristics of typical applicants that apply to particular residency programs. You can find this information at: http://www.nrmp.org/data/chartingoutcomes2011.pdf.

## Overcome Blemishes

The USMLE Step 1 Exam has served as a new beginning for countless students needing a fresh start. Whether that blemish is poor first-year grades, time off during your medical career, or a lack of research/community service, the Step 1 score often will get your figurative "foot in the door," long enough to explain your special circumstances. The Step 1 Exam has the amazing power to overcome all blemishes on your residency application because it shows that you have the ability to excel.

# Chapter 4

## Content Strategy

This section of the Method answers the question: What do I need to know for the Boards? With over 500 topics covered on the exam, students get overwhelmed by the immensity of the possible fund of knowledge. Students consistently rack their brains with the eternal question: Do I really need to know this for the exam? With this constant uncertainty, it is impossible to focus on key information and connect major ideas.

For this reason, we created the *Step 1 Method* Framework. The framework is a guide that helps students focus on what is important within the extensive subject matter in the basic medical sciences. The framework is derived from determining how the NBME thinks about information on this exam. Naturally, the way the NBME thinks about information translates into the way it asks questions. Thus, if you cover and understand the framework of a particular topic, you can feel confident that you will be able to answer all the potential questions that the examiners can raise.

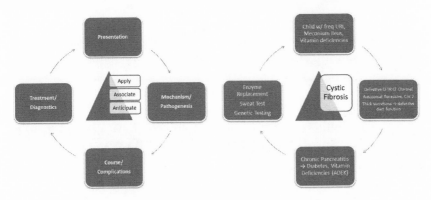

*Figure 1 The Step 1 Method Framework*

This graphic illustrates the *Step 1 Method* Framework and how it is meant to encompass the key factors you need to know about each topic. The framework pictured is for pathology/pathogenesis topics, but it can be appropriated for topics in each discipline or category. We are developing new frameworks for each subject as appropriate (e.g. biochemistry, pharmacology, etc.). Please visit our website for all of the up-to-date frameworks. There are four to five aspects of every topic the Board examiners expect you to know; these areas are where their questions come from. By focusing on these areas in anticipation of obtaining actual questions on them, you will be prepared for a great majority of questions on the USMLE Exam.

1) **Presentation**: Learn the clinical presentation of the disease. How will the patient present to the hospital or clinic? What signs (these are the disease manifestations that the doctor will see in the patient's physical exam) and symptoms (these are what the patient will complain of) will the patient have? The NBME will not tell you what disease process the patient has; they will expect you to make the mental "step" to connect the presentation with the disease process. This is the way the majority of clinical vignette USMLE Board questions will present.

> Q: A 55-year-old man with a history of coronary artery disease, hypertension, COPD, and hyperlipidemia comes into the clinic complaining of increasing shortness of breath and cough. On physical exam you observe a jugulovenous pulse of 10 cm $H_2O$ and 2+ pitting lower extremity edema. With what are these signs and symptoms consistent?
>
> A: Congestive heart failure

- If we were to make the previous question a three-step question, we would begin with the preceding presentation and ask:

> Q: In the preceding example, what pathophysiology would one see on tissue biopsy of the patient's lungs causing the current episode of shortness of breath?
>
> A: Increased hydrostatic pressure causing congested vascular beds and alveoli (cardiogenic pulmonary edema)

- In this way, the first step was to identify that the patient's exam was consistent with congestive heart failure. Then, for the second step you would have to determine that the lung symptoms in congestive heart failure are due to pulmonary edema. Finally, for the third step you would have to know that the pathophysiology behind pulmonary edema involves increased hydrostatic pressure causing congested vascular beds and alveoli. A likely wrong answer choice in the previous question would describe the pathophysiology of COPD. Given the patient's history of COPD, it is possible that he could have this pathophysiology, but the NBME asked for the "current cause" of shortness of breath.

2) **Mechanism/Pathogenesis**: Learn the molecular basis of the disease. Is it an enzyme deficiency, over-activity, or abnormal activity? Is it a protein deficiency or malfunction? Make sure you can name the molecule/enzyme of importance, what system it is in, and the intersection points with other frameworks.

   Learn the pathophysiology of the mechanism of the disease process. How does the abnormal mechanism actually cause a disease that produces signs and symptoms? What happens when the receptor doesn't bind the necessary protein or the enzyme's deficiency leads to a build-up of a certain product? Usually, it is as simple as there is too much of something, or there is not enough of something. Imbalance in the homeostasis of life is the basic cause of disease. Always try to connect pathophysiologic processes with clinical signs/symptoms as this is a common source of NBME questions as seen in the previous example.

3) **Course and Complications**: Understandably, a disease itself is a complication of some abnormality in the homeostasis of human physiology. But as you will find, the same disease will come in many varieties of severity and decompensation. Certain diseases, such as diabetes, will begin in one organ system (endocrine) but will worsen such that they affect several other systems (cardiovascular, renal, neurologic, and ophthalmologic). You must understand the natural course of a disease and how it can progress to an end-stage state. You must essentially know

the worst-case scenario for the disease. For diabetes, it would be coronary artery disease (CAD) and myocardial infarction (MI), end-stage renal disease (ESRD), diabetic neuropathy, and diabetic retinopathy, respectively. In addition, there are a number of disease processes that will result in macroscopic complications for the patient that were not a part of the original disease process. These complications are usually a result of the worsening of the disease process. For instance, blindness is one complication of the worsening of diabetic retinopathy. Another example of such a complication is the rupture of the free wall of the heart days after a myocardial infarction.

4)    **Diagnostics**/Treatment: This area of the framework is tested least on the USMLE Step 1 (i.e. this is more of a focus on Step 2), but in recent years, the NBME has expected students to know major diagnostic tests and in some cases, definitive treatment for certain conditions. You must know if there is an easy diagnostic test for a condition [e.g. sweat test for cystic fibrosis, or cardiac enzymes (Troponins, CK-MB) for myocardial infarction].

You must be able to know the answers to the following questions: Is this a treatable condition? Can you treat it with medicines, or does it require surgery? Is it an autoimmune condition in which you must suppress the immune system, or is the cause unknown? At this stage in the game (i.e. Step 1), you do not need to know all definitive therapies, but you must at least know the category of the first-line therapy if there is one. For example, if a patient presents with symptoms of hypothyroidism, you will have to be able to treat that patient with levothyroxine or synthroid. Another clear-cut example is that of minimal change disease; the amazingly effective treatment is steroids. More often than not, you will encounter cases where the patient was treated for a known condition and you must be able to determine the mechanism through which the medicine works and what potential side effects the patient could have from this medicine.

- If the treatment is surgical, what type of surgery will they have? Will it be a partial removal of the organ in question or a complete excision? More importantly, what are the potential effects of the surgery? A classic example is the following:

Q: A patient was diagnosed as having thyroid cancer or an atypical thyroid nodule. The definitive treatment is excision of part or all of the thyroid gland. The patient undergoes the standard therapy and presents to the clinic two weeks later with complaints of cramping and muscular twitching (or worse, abnormal palpitations). What happened?

A: Hypocalcemia due to hypoparathyroidism, due to inadvertent removal of the parathyroid glands and lack of endogenous parathyroid hormone (PTH).

- In this way, the Board writers will connect normal therapies with abnormal results and complications, and you must be able to anticipate them and spot them when they arise.

This outline of the different characteristics of disease can be applied to all topics and should serve as a framework for determining whether or not you have mastered a particular topic. You can apply the same criteria to microbiology and pharmacology with their respective disease processes caused and treated, respectively. It is helpful to use this framework when approaching several diseases in a stepwise fashion to avoid the feeling of being overwhelmed.

# The 3 A's: What to Do with the *Step 1 Method* Frameworks

The graphic within the framework diagrams points out the 3 A's: Apply, Associate, and Anticipate.

Apply: We want you to apply the framework to every topic you encounter because it is a great way to ask yourself if you really know the topic. You should be able to relate the connections of all of the elements without looking at your notes. This will be the scenario on the exam. If you apply the Method, studying for the Step 1 Exam will encompass the last half of your second year of med school. You will find yourself out of school, and COPD will pop into your head. You should ask yourself, "Do I know the framework for COPD?" You should be able to go through it. If not, don't get discouraged. Just look it up in your central text (See Study Strategy for more on the central text) the next time it is convenient to do so.

Associate: Each element of the framework connects with another. Understand the signs and symptoms caused by pathogenesis. Understand how the pathogenesis causes the course of the disease and other complications. Understand how potential treatments alleviate the causative pathogenesis or mechanism. Further, understand how different frameworks and topics can be associated with each other. In our previous example, cystic fibrosis can lead to pancreatitis, vitamin deficiencies, and diabetes, each of which has its own framework. The NBME loves it when distinct frameworks coincide because it can test two topics at once. Further, as we discussed earlier, the NBME needs to have questions with differing levels of difficulty. Questions that include associations among multiple frameworks require a more complex and deeper understanding than most students are prepared for.

Anticipate: The ultimate goal of our content strategy is for students to begin to think like the NBME. When you understand how the examiners think about information, you will be able to anticipate potential question topics. Most students only focus on memorizing details and numerous facts. As you're learning information, you should always be asking yourself: If I was the NBME, how would I craft a question about this topic? Be constantly thinking in two and three steps, tying together ideas across frameworks.

# A Deeper Dive into Three-Step Questions

Over the past decade, the USMLE Step 1 Exam has evolved. The exam used to focus on individuals facts and details. You can still see some of these old questions on older NBME CBSSAs. The NBME has shifted to an exam that is more clinically oriented. The exam tests a student's ability to synthesize complex information and integrate across topic disciplines. The quintessential Board question is called the "three-step question." The examiners will present a clinical scenario and describe the signs and/or symptoms of a patient without giving you the diagnosis. The actual question will query about the side effect of a medication used to treat the patient's condition or the mechanism/pathogenesis of a complication of the patient's disease. In order to answer these types of questions, you are required to make three mental "steps" or assertions prior to arriving at the answer.

The following are some examples of three-step questions utilized by the Boards:

Q: A 55-year-old male with several risk factors for cardiac disease presents to the emergency department with chest pain. Among other medications, he is given an aspirin, a nitrate, supplemental oxygen, and a beta-blocking agent. Notably, he is not on any of these medicines at home. Shortly after receiving these therapies, he becomes acutely short of breath. On auscultation of his lungs, you hear diffuse wheezing. Physical examination reveals a barrel chest and a chest X-ray shows a flattened diaphragm and hyperinflated lungs. On further questioning you find that he has a 50-pack/year history of tobacco use. What is the pathophysiology of his shortness of breath?

A: The first step is realizing that the patient (who has a history of smoking, barrel chest, and hyperinflated lung fields on CXR) also likely has COPD (chronic obstructive pulmonary disease). The second step is realizing that with his acute shortness of breath and wheezing, he is likely having a COPD exacerbation. The third and last step is realizing that the beta-blocking agents he received for his chest pain can cause shortness of breath in COPD and asthma patients by blockade of B2 receptors on bronchiole smooth muscle thus preventing bronchodilation and causing wheezing.

Interestingly, although the above question began with the presentation of chest pain, the question was not about cardiac disease. Many students will be distracted by unnecessary details in questions as they read through questions. The NBME does this on purpose to determine if students can sift through unnecessary detail. In our USMLE Strategy section, we cover how to read USMLE questions effectively to prevent being sidetracked by distractors.

Here's another example of a three-step question:

Q: A 54-year-old female with diabetes mellitus and hypertension presents to you as her new primary care provider. You notice that her blood pressure is uncontrolled on two agents: lisinopril and amlodipine. You increase the doses of both medications and see her in follow-up two weeks later. Her blood pressure is improved, but she complains of vague and diffuse muscle weakness and palpitations. You order a chemistry panel but before that returns, you also order an EKG which shows a prolonged QRS segment and peaked T waves. What is the mechanism of the patient's new symptoms?

A: The first step is to understand that lisinopril belongs to the ACE inhibitor class of anti-hypertensives and amlodipine is a calcium channel blocker. The second step is realizing that the patient is presenting with symptoms of hyperkalemia (muscle weakness and associated EKG findings). The third step is to understand that one of the main side effects of ACE inhibitors is hyperkalemia due to inhibition of the activation of angiotensin II and therefore decreased production of aldosterone and decreased potassium excretion.

The NBME examiners love to start out a question with a pathological intro and turn it into a pharmacology or biochemistry question as just shown. They only have 350 questions to test over thousands of permutations of the 500+ topics on the test. The way they are able to do this is by integrating disciplines and connecting remote ideas.

The diagram for the *Step 1 Method* Framework suggests that all the elements of the framework are connected. We want you to understand the spectrum of information on a topic and how it is related. The signs and symptoms (Presentation) of a disease process are caused by the mechanism and pathogenesis of the disease process. Many times, students will just memorize the signs and symptoms of disease; when the Board examiners ask about the pathogenesis, the student cannot answer the question.

These examples show that it is very important to take all concepts and facts down to the mechanistic level. It is not enough to know *what* something is, but rather, you must know *why* something is. Most students are very diligent and will memorize large amounts of information and questions they saw in the Qbank. These students are at a loss when the NBME creates a hypothetical scenario such as a laboratory experiment. The topic in question will have the same mechanism and pathogenesis as a topic with which they are very familiar, but they are unable to abstract the information they know. The key to success is to understand how the elements of the framework connect and then be able to apply that information.

In the illustrated example of the *Step 1 Method* Framework, we have shown you what the framework for cystic fibrosis looks like. You can see that we don't get into too much detail, but we do cover the main ideas for each element of the framework. On a day-to-day basis as you study, you should ask yourself if you know the framework for a particular topic. In our Study Strategy section, we show you how you should incorporate the framework when working questions.

# The Concept of the "One-Liner" and "Finding the Stem"

When thinking about USMLE style "three-step" questions, usually the "first step" that is required is to identify the disease process or topic described in the question. As explained, the NBME will not directly tell you that a patient has cystic fibrosis; they will describe the signs and symptoms that patient presents with and expect you to know that the patient has CF. Thus, this "first step" is usually encompassed by the "Presentation" element of the *Step 1 Method* Framework. This first step is crucial because if you assume the NBME is referring to a different disease process or topic, you will be answering the wrong question. Thus, both accuracy and speed are important in making this first step.

To train students how to increase their accuracy and speed in making the "first step," we focus on two exercises called "finding the stem" and building "one-liners." Often, the signs and symptoms of a disease or topic can be described in a short, incomplete sentence called "the one-liner." When learning topics, you should construct one-liners to help you envision which key information you will be looking for in questions. In contrast, the

NBME will take an eight- to nine-line question to adequately layout an entire clinical presentation or biological scenario. The "one-liner" is often sitting within this long question, hidden amongst unnecessary detail and distracting information. We call this hidden one-liner the "stem" of the question. Thus, the two terms are synonymous. The "one-liner" is built by the student when trying to tie together the most important factors of a topic, while the "stem" is the one-liner created by the NBME and buried in distracting information. Thus, every time a student works a question, their goal is to "find the stem" so that they can take the first step to answering the question.

Typically, the process of building a one-liner comprises two main components: First, who is the patient? What are their demographics? Are they characteristically male or female? Young or old? African-American, Asian, or Caucasian? Second, what are the main signs and symptoms of the disease? When you combine those two main elements, you have built a one-liner.

In the example of cystic fibrosis, a sample one-liner would be, "Young patient with history of upper respiratory tract infections, vitamin deficiencies, and meconium ileus in infancy." When encountering a question about a patient with cystic fibrosis, there may be significant detail about the patient and history, but they may only mention that the patient had meconium ileus in infancy and has newly diagnosed malabsorption. Although in this "stem" the NBME did not give us all the prototypical symptoms of cystic fibrosis, the patient still most likely has CF, and we should continue with our thought progression down the CF framework to answer the question. A common saying in clinical practice is, "Patients don't read the textbook." This refers to the fact that patients won't always have all the symptoms that are associated with their disease. Further, this is a way that the NBME can make questions more difficult. By inserting only two rare symptoms of a condition and leaving out those most commonly memorized by students, the NBME can guarantee that 20–30 percent of students will answer the question incorrectly.

As always, visit our online course to view the latest videos, frameworks, and other multi-media in our Content Strategy section.

# Section 2

# USMLE Strategy

# Chapter 5

## The Anatomy of a Question

The USMLE Step 1 Exam is written in a unique and characteristic style. The NBME trains its staff to write questions in a rigorous, uniform format. If students learn this format, they will greatly increase their accuracy in answering questions correctly. In this section, we cover strategic issues that are unique to taking the USMLE Step 1 Exam.

## Understanding Content – The USMLE Style of Questions

USMLE Step 1 questions are written to contain a lot of information, most of which is not essential to answering the questions. Students often lose their focus when reading these information-rich questions. A common complaint is that students finish reading a very long question and still don't know what the NBME is asking. They find themselves attempting to answer the wrong question, which inevitably leads to the wrong answer. The root problem is that students do not understand the content of the question.

To remedy this, we teach students how USMLE questions are written. The NBME examiners have a question-writing manual in which they teach new Boards examiners how to write questions. In our Anatomy of the Question Module, we show you how to make this knowledge work for you. If you know the basic structure of the question, you will know where to look for the desired data.

In the following diagrams, you can see the schematic of USMLE Step 1 questions, and this schematic is illustrated within an actual NBME Step 1 question.

# The "Body and Lead-In"

These are the two main parts of the text of the question. The "body" is all of the detail of the question, and the "lead-in" is the actual question portion of the larger "question." Many students get lost in the body of the question. By the time they reach the lead-in and figure out what the question is actually about, they have forgotten many of the details in the body. So, they have to go back to the body and read it again now that they know what they are looking for. To prevent this common time-wasting scenario, we recommend that students glance at the lead-in first. By doing this, you will have two valuable pieces of information: First, you will know what element of the framework the NBME is looking for. Is it a mechanism/pathogenesis question? Diagnosis/Treatment question?

Second, often the NBME will also include the physiological system in the lead-in. As in the following example, "Pathological examination of his brain will disclose an abnormality in which of the following?" the NBME has just told you, "This is a neurological pathophysiology question: Good luck." At that moment, you can put on your neuropath "hat" and ask yourself: How well do I know neuro path? Your mind is like a library: telling your brain where to look for the stored information allows it to reference associations and improves information retrieval immensely. Further, by knowing these two key pieces of information up front, you will be able to go through the body and pick out the important information to build your one-liner.

# "The Identifier"

The "identifier" tells you demographic information about the patient. The NBME will give you data about the age, sex, ethnicity, occupation, and other important past medical history about the patient. The reason you are given identifying information about the patient is to clue you into the risk factors the patient has for a particular set of disease processes. Many times students overlook these clues, but in clinical practice, the background of the patient is just as important as their presenting signs and symptoms.

In the following question, you are told the patient is a "50-year-old man with a history of alcoholism." To continue with our sequential process of working the question, we know that the question is a neuro path question from reading the lead-in first. So now, we think: What neuro path processes do I know that are associated with alcoholism? Some of you may be drawing a blank. That's okay—you're reading this book early and have plenty of time to start learning information on the level required by the NBME. Some of you automatically start to think of Wernicke-Korsakoff's syndrome. This is a good thought. Another neuro path condition to think of with respect to alcoholism is hepatic encephalopathy, which, along with the other stigmata of end-stage liver disease (spider angiomas, palmar erythema, caput medusa, esophageal varices), can also present with asterixes, a neurologic phenomenon.

It is important not to go too far when building your one-liner. A mistake many students make is coming to a conclusion too soon. Some students may have put together alcoholic and neuro path, already assuming Wernicke-Korsakoff, and will jump to the answer choices. Or worse, they will assume it is Wernicke-Korsakoff and read the remainder of the question without testing their assumptions by fact-checking the key signs and symptoms.

# "Key Associations"

After placing an identifier, the next portion of the question in the body will be detail about the patient or topic in question. Your goal is to pick out the "key associations" that will enable you to build your one-liner. These key associations are the signs and symptoms associated with the disease process or topic in question. To do this, you must be able to pick out details that are unique to the topic in question and separate them from other similar disease processes. This is where knowledge of different topic frameworks comes in handy. The most difficult questions will describe a syndrome that has several similar syndromes; the NBME will then choose wrong answer choices that pertain to these similar syndromes. If a student assumes the wrong syndrome, there will be an answer choice that baits them into picking it.

A majority of the detail in the question will be unnecessary. For example, most normal findings, such as normal vital signs and anecdotal information, won't help you build a one-liner. Non-specific symptoms that are shared in many diseases, such as arthritis or fatigue, are also not very helpful. You are looking for abnormal and unique symptoms. Further, you only need three or four signs and symptoms in constellation to build your one-liner and make a diagnosis. Our fund of knowledge is finite; sometimes we may only know two signs/symptoms of a particular topic, but the question may list five or six. In these situations, you must go for it. It is impossible to know everything, and if your best guess is based on limited information, it is better than no information at all. Sometimes students make the mistake of talking themselves out of an answer because of what they don't know. They worry that the four symptoms/signs that they are not sure of will make that answer wrong; usually, this is not the case. You should put in the hard work up front, learn the information the right way, and then operate from a position of confidence based on the information that you do know.

The following question states that the patient has difficulty with short-term memory. This is a great key association. Next, the NBME examiners provide some unnecessary detail. They tell you, "He is unable to recall the date and cannot remember what he ate for breakfast this morning." This is not new information; they are simply repeating the idea that he has trouble with short-term memory. Sometimes, this new information will confuse students. They will postulate as to whether there is something special about him forgetting something this morning or the fact that he can't remember the date. In most cases, try and keep it simple: you are looking for key signs and symptoms similar to the ones you built one-liners for when you learned about the topics the first time.

Next, the NBME describes the situation where the patient believes something is not true. To some students, this may seem like unnecessary detail. But to others, they know that this is an example of a symptom/sign called "confabulation." Fortunately, the NBME is friendly to students when it uses "pathognomonic" signs/symptoms in a question. These are signs/symptoms that are largely associated with only one disease or topic. Unfortunately, pathognomonic symptoms help us only if we know what they are. In this case, "confabulation" in an alcoholic is pathognomonic for a certain disease state.

We should be adding key associations as we go, building a strong one-liner as we read through the question. As we encounter new symptoms, we will start to get confident as to whether we know what topic the NBME is describing. Once we've come to an assumption, we should fact-check new signs/symptoms against our one-liner and assumption to make sure they are compatible. As mentioned previously, if we encounter a new sign/symptom that we are not sure of, or have never heard of, we should not negate a strong one-liner or assumption just because of this new info. Thus, at this point, we have enough information to build a strong one-liner: an alcoholic with short-term memory loss and confabulation.

# "Unnecessary Detail"

Now that we have built a presumptive one-liner, we should continue reading and fact-checking our assumption. The next line tells you, "His long-term memory appears intact." This is not new information that will help us make a new diagnosis or negate our presumptive diagnosis. The next line is very interesting. They tell you, "The patient dies shortly thereafter of a myocardial infarction." This is something that the NBME examiners love to do. They love to give you extra information that can be true and, at the same time, unnecessary and distracting. There are several reasons why this patient could have had a heart attack. Had you not read the lead-in first and known they were asking you about a neuro path topic, this last detail would have likely thrown you off. You would have likely been taken aback as the prior symptoms only had to deal with memory problems. You would have quickly started to think about what disorders cause heart attacks and memory problems. Unfortunately, that would have taken you down a road that would likely lead away from the right answer.

In this case, the unnecessary detail is relatively benign and unhelpful; these extra details can sometimes purposefully distract and fool students into picking a wrong answer choice. There are many questions where the NBME will provide unnecessary detail and also provide answer choices that correspond with that piece of unnecessary detail. These occurrences lead to the common notion that the NBME tries to "trick" students. This fits with the oppositional nature that most students have with this exam and the NBME. I would like to push back on this notion. We explained earlier that the NBME attempts to probe the depth of a student's understanding. If students are easily distracted and confused by unnecessary detail, then the student did not completely understand the concept/topic being tested. By modulating the amount of distracting information, the NBME can create the difficulty spectrum of questions required on the USMLE Step 1 Exam.

# "Taking the Steps to Get to the Answer"

After having built a one-liner that stood up after finishing the body of the question, you return to the lead-in. You now know which element of the topic's framework they are asking for. Before you can take the steps along the framework, you have to first determine what topic they are testing you on. Most commonly, the NBME examiners have described the Presentation, and it is up to you make the diagnosis.

In this case, our one-liner was "alcoholic with short-term memory loss and confabulation." This is the presentation of Korsakoff's syndrome. Most students spend a majority of their time memorizing one-liners and mechanisms. Most students may know that the mechanism of Wernicke-Korsakoff's syndrome is due to the thiamine deficiency usually seen in alcoholics. What most students don't do is learn the entire framework of disease topics. What you quickly see is the NBME asking for the pathogenesis of Korsakoff's syndrome. Simply knowing that thiamine deficiency was the cause was not enough. This is the common scenario students find themselves in with the USMLE Step 1 Exam. They have exposure to many topics tested, but they find that their knowledge base does not go deep enough.

In order to get to the answer, we have to walk through the steps in order. The first step was knowing that the patient had Korsakoff's syndrome. The second step was knowing that the mechanism of Korsakoff's syndrome is thiamine deficiency. The third and final step in this question was knowing that the pathogenesis of thiamine deficiency in the central nervous system leads to mammillary body degradation.

When working questions in your daily study, we teach you not to stop with the framework element that was asked in the question. We want you to work through the entire framework. The NBME does not write completely new questions every year; it recycles old questions and changes them such that they will ask about a different framework element of the topic. In this case, the NBME could keep the same body and ask "What diagnostic test would confirm the patient's condition?" or "What would be the best initial treatment for the patient?" The answers to these questions would revolve around thiamine (i.e. thiamine assay and IV thiamine, respectively).

For more worked-out examples, go to the online course and view "The Anatomy of a Question" modules. Ultimately, reading questions in this fashion will become second nature, and you will do it automatically without thinking. When working questions, we don't want you to write out one-liners and diagram the questions as we've done in the examples. We want this style of reading questions to become a part of your armamentarium in understanding questions and increasing your accuracy of answering questions.

# The Anatomy of a Question

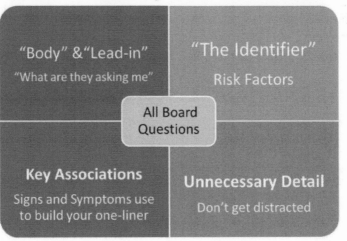

*Figure 2 Anatomy of the Question Schematic*

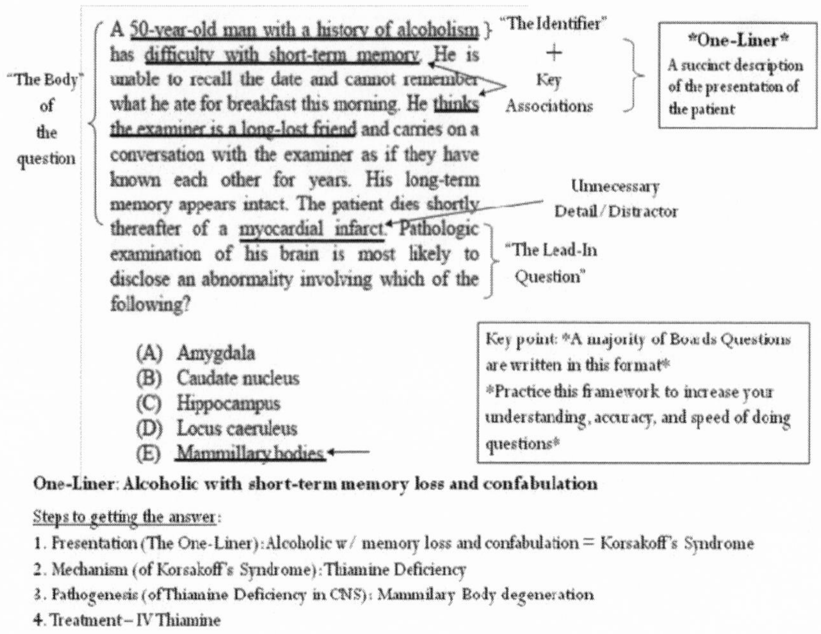

"The Body" of the question

A 50-year-old man with a history of alcoholism has difficulty with short-term memory. He is unable to recall the date and cannot remember what he ate for breakfast this morning. He thinks the examiner is a long-lost friend and carries on a conversation with the examiner as if they have known each other for years. His long-term memory appears intact. The patient dies shortly thereafter of a myocardial infarct. Pathologic examination of his brain is most likely to disclose an abnormality involving which of the following?

"The Identifier" + Key Associations

**\*One-Liner\***
A succinct description of the presentation of the patient

Unnecessary Detail / Distractor

"The Lead-In Question"

(A) Amygdala
(B) Caudate nucleus
(C) Hippocampus
(D) Locus caeruleus
(E) Mammillary bodies

Key point: *A majority of Boards Questions are written in this format*
*Practice this framework to increase your understanding, accuracy, and speed of doing questions*

One-Liner: Alcoholic with short-term memory loss and confabulation

Steps to getting the answer:

1. Presentation (The One-Liner): Alcoholic w/ memory loss and confabulation = Korsakoff's Syndrome
2. Mechanism (of Korsakoff's Syndrome): Thiamine Deficiency
3. Pathogenesis (of Thiamine Deficiency in CNS): Mammillary Body degeneration
4. Treatment – IV Thiamine

*Figure 3 Anatomy of the Question Principles in action*

# Chapter 6

## Question Strategy

A key principle of the *Step 1 Method* is that you should do everything you can to prepare for the exam prior to walking into the test. You should practice the study and preparation strategies and acquire a deep understanding of the basic sciences on the exam. On the other hand, you should also expect to encounter material you have never seen before. You should expect that you will not know the answer to every question and will need to figure out some things on the fly. For these reasons, a major component of our program is Question Strategy; these are all strategies to use when you don't know the answer to the question. Students commonly refer to these "black arts" as "test-taking skills." If students do not feel they have these skills, they often dub themselves a "bad test-taker."

To be clear, we do not believe the USMLE Step 1 Exam can be gamed and strategized completely. Test-taking strategies on the Step 1 Exam only work if you have adequately prepared before walking into the exam. In this chapter, we examine general test-taking strategies that can be applied specifically to the style of the USMLE Step 1 Exam. After learning these strategies, you should always have an attack plan no matter how difficult, long, or confusing the question. Having an attack plan prevents you from getting psyched out and frustrated. By remaining calm and confident, you will be able apply the information you've spent months studying.

## The "Bad Test-Taker"

In developing the *Step 1 Method* and working closely with our partner schools, I've spent a majority of my time with students who have termed themselves "bad test-takers." When we work with client schools, we prospectively identify students who may have test-taking difficulty, which can be evidenced by test scores and self-identification. In general, students who show signs of test-taking difficulty fall into three major camps of varying preparation level and test-taking anxiety/skills. I believe we can learn the most from students who are adequately prepared but underperform on exams. If you talk with the student before or after the exam, they can eloquently relate a deep level of understanding on tested topics, but in the moment of the exam, a perfect storm of negative psychological phenomenon occur.

In general, my experience has taught me the major difference between "good" and "bad" test-takers is the ability to access available knowledge once the test-taker knows that they do not know the answer to the question. The "good" test-taker remains confident despite understanding that they do not know the answer. They develop strategies to utilize information that they do know about peripheral topics, the answer choices, and the language of the question in order to make a best guess. Throughout this process, they are calm and even eager to make a best guess. When you query them after the fact, they are proud of their deductive reasoning skills, even if the logic is faulty or based on false assumptions. This observation in USMLE Step 1 prep is backed up by cognitive science research on confidence and test taking. Correcting for preparation and ability levels, more confident test-takers have a higher accuracy rate in their "guessing" that correlates to their confidence levels.[2]

In contrast, when self-termed "bad" test-takers realize that they do not know the answer to a test item, they shut down mentally. Their anxiety levels rise, and they slowly begin to focus on feelings of dread and fear about missing the question. Their immediate reaction is to read the question again, hoping they may learn something new or find something they missed. After returning to the same conclusion that they do not know the answer, they start to go through a process of deduction I call "find the best-looking answer." Instead of focusing on what they know about the

---

[2]    Stankov, Lazar; Crawford, John D. Self-confidence and performance on tests of cognitive abilities. Intelligence. September 01, 1997.

topic at hand or about the answer choices, they search for an answer that "looks right." They will gravitate toward answers that contain the same phrasing as phrases in the question. Unfortunately, this plays right into the NBME's distractors' hands. The NBME purposefully puts in distractors that "look" right but are factually wrong. By making decisions based on appearances and not principles, these "bad" test-takers will consistently guess wrong.

In the event that the "bad" test-taker has an inclination based on principles, they will be racked with indecisiveness. They often talk themselves out of this more correct answer in favor of an answer that "looks" better. To make matters worse, the negative energy and demotivation that occur when encountering a difficult question affects the psychology of the following questions. In order to overcome these challenges, the "bad" test-taker must improve their deduction using information available to them. They must combat indecisiveness when they have an inclination based on factual information.

The psychology of terming oneself a "bad" test-taker is self-destructive. It breeds unnecessary self-doubt and indecisiveness. A student should be confident they've prepared to the best of their ability. The test-taking strategies used by "good" test-takers can be learned. Combining optimal preparation and practice with the implementation of these strategies will give students the best chance for success. Students who have had testing difficulty in the past should wipe the slate clean and do their best to reach their goals. Using the *Step 1 Method*, you will know exactly where your performance stands throughout your Intensive Study Period.

# Running out of Time

One of the most common problems many students have on the exam is running out of time. There is no bigger forfeiture of points than leaving unanswered questions at the end of every block. Because there is no penalty for guessing, you are giving away free points. Further, you will find that the NBME will place very straightforward questions at the end of every block. These questions are often disguised in text that is intimidating because of its length; but if students have the time to read the question, they will likely answer it correctly. The common scenario is that many students spend a majority of their time on questions 1–40 and then rush through the last six, picking answers that "look right."

In order to finish each block of questions on time, students must have a prepared skipping strategy. Many medical students scoff at the concept of skipping questions because they have never had to skip questions before. Each student on average has 1 minute and 18 seconds per question. It is natural for many students to spend five to six minutes on the toughest questions. Ironically, because of their difficulty, these are the questions students have the highest probability of getting wrong. Thus, does it make sense to spend the most time on questions you have the highest probability of getting wrong? By utilizing a calculated skipping strategy, we teach students to spend a majority of their time on questions they have the highest likelihood of getting right. Once you have ensured correct answers on these questions, you can spend extra time on the most confusing questions on the test.

## The Skipping Strategy

Every student has their own areas of weakness and individual types of questions that give them fits. Some students do not like experiment-based questions, while other students cannot stand epi-biostats question. Thus, the questions that students choose to skip will be personal. But when you choose to skip a question, the format should always be the same.

You should make sure that you read all questions effectively using the Anatomy of the Question principles. After reading a question, you come to one of three conclusions fairly rapidly. You either:

1.   Know the answer to the question;
2.   Know you can figure it out if you go through a few mental steps; or
3.   You have no idea what the answer is.

If you find yourself in the first two categories, you should remain on the question and answer it to the best of your ability even if it takes you a few minutes. If you find yourself in the third category, you should practice the skipping strategy.

The skipping strategy aims to correct two major errors students make when attempting difficult questions. Often when students are faced with difficult questions, their first instinct is to read the usually long question two and three times. They assume that they missed something and that by reading the question again, something magical will happen. On the most difficult questions, this wastes another two to three minutes. Often after reading a question multiple times, a student is racked with uncertainty and cannot come to an answer. Instead of marking their best guess, the uncertainty will prompt them to mark and skip the question, moving on to the next question, vowing to return. If the student does have time to return to the question, they will have to read it all over again, go through the same thought process, and still remain stumped. They will often end up at the same best guess, but this belabored process took eight minutes instead of two minutes.

In order to practice the skipping strategy correctly, you must read questions effectively the first time using the "Anatomy of a Question" principles. When you have decided that you don't know the answer and plan to skip it, look at the answer choices and make a best guess. Go with your instincts and come up with a strategy to answer the question. Don't put pressure on yourself. If you don't know it, you don't know it. It's okay. Make a best guess, mark the question, and move on. This process should take you no more than two minutes. You should expect to skip at least five questions per block. If you consistently use this skipping strategy on the most difficult questions, you will find yourself with plenty of time at the end of the block to return to these marked questions. You can reread the question, and if you have remembered a new piece of information, it's okay to change your answer. (We cover the science of changing answer choices later in this chapter.)

This skipping strategy is very powerful in reducing the time-related anxiety on the test. Students who are well prepared and know what they know and know what they don't know, find themselves with 5–10 minutes at the end of every block to review marked questions. We have worked with several students who had to take the MCAT multiple times because of severe time-anxiety issues. These students have successfully implemented the skipping strategy with great results.

Another benefit is that the skipping strategy enables you to use your prep and study time for learning information instead of trying to gain speed. A common mistake students make is to try and time themselves when completing blocks of questions during their study period. As we'll cover in the Study Strategy section, you cannot expect to learn information at a deep level if you are constantly speeding through questions. By waiting to evaluate yourself during practice tests, you can use the question banks as educational tools that help you learn the information.

# Fear of Standardized Tests

Another hurdle for many students facing the USMLE Step 1 is a generalized fear of standardized tests. Interestingly, these students rarely have problems with their school exams, but when it comes to national standardized exams, their anxiety levels rise and their confidence drops. This phenomenon is usually due to a poor experience with the MCAT exam. Many med students suffer through the MCAT and/or take it multiple times. When these students get to medical school, they feel that they've escaped that scarring experience. It's no surprise that those familiar trepidations return when Step 1 season begins.

Fortunately, the MCAT and the USMLE Step 1 are very different exams. The MCAT is as much a reading comprehension test as it is a test of depth of knowledge. The AAMC has published data showing that humanities majors score better on the MCAT than do biological science majors (Avg. Matriculant Score 31.7 v. 30.9).[3] The MCAT is predominantly text-heavy passages, and a majority of questions are based on passage reading. Despite a deep understanding of the science behind a question, a student can easily get tripped up on the author's phrasing. Conversely, the MCAT is an exam in which the author's tone can lead an astute test-taker to the answer without significant understanding of the science behind the question.

---

[3]    MCAT and GPAs for Applicants and Matriculants to U.S. Medical Schools by Primary Undergraduate Major, Association of American Medical Colleges. https://www.aamc.org/data/facts/applicantmatriculant/. 2011.

The USMLE Step 1 Exam, on the other hand, is an exam heavily weighted in basic science content. It is not an exam that can be purely strategized or gamed without a deep level of understanding. The NBME will often include distractors in questions to fool those would-be gamers. Most recently, the NBME has started to include seven and eight answer choices in questions to eliminate the 20 percent chance of guessing correctly in the standard five-answer-choice system. "Three-step questions" and mechanism/pathogenesis answer choices make a superficial understanding of basic sciences insignificant in answering questions.

While the MCAT is a good measure of test-taking ability, the USMLE Step 1 is an exam that primarily measures your fund of knowledge and your test-taking ability second. Thus, to excel on the USMLE Step 1 Exam, you must prepare both for the content and the exam itself. If done correctly, you can erase any past standardized exam stumbles. To illustrate this point, we track our students' MCAT exam, NBME practice exam, and Step 1 Exam performance. We plot these and a number of other variables to determine if there is a correlation among the different data points. Interestingly, we find data in our student population that varies slightly from the norm. There have been several published studies that show a correlation between the MCAT and the USMLE Step 1 Exam. A widely cited study by Tyrone, et al. from 2007 published an r value of 0.6.[4] In 2010, Morrison, et al. repeated a study correlating the CBSSA (an NBME practice test) with USMLE Step 1 performance. In this study, the authors found an r value of 0.67.[5]

In our data set, we found a slightly different pattern. In a single data set, we compared the first NBME practice exam taken, MCAT score, and USMLE Step 1 Exam score. We found that the correlation between the first NBME practice test and the USMLE

[4]    Tyrone, et al. The predictive validity of the MCAT for medical school performance and medical board licensing examinations: a meta-analysis of the published research. Acad Med. 2007 Jan; 82(1):100-6.

[5]    Morrison, et al. Relationship between performance on the NBME Comprehensive Basic Sciences Self-Assessment and USMLE Step 1 for U.S. and Canadian medical school students. Acad Med. 2010 Oct; 85(10 Suppl): S98-101.

Step 1 Exam was 0.76, while the correlation between the MCAT and the USMLE Step 1 Exam was only 0.34. This 0.34 value is much lower than what has been reported by other authors. When you take a closer look at the raw data, you see students with 31s and 32s on the MCAT (the current national average of matriculants) scoring high 250s and 260s on the USMLE Step 1 Exam. These USMLE Step 1 scores are presumably in the top fifth percentile in the nation. (Note: The NBME does not release percentile scoring, but one can reconstruct a distribution based on national mean and pass rates.) These students from the data set are far outperforming what their expected USMLE Step 1 score is based on their MCAT score.

This trend shows a powerful theme that we have known empirically: the MCAT has low predictive ability for the USMLE Step 1 Exam when students prepare for the Step 1 Exam in an effective way. Admittedly, our data sets are smaller, and we are still aggregating the following data, and performing more stringent statistical analyses on the sets in the hopes of publishing it. But this is something we have noticed year after year: motivated students who score in the 20s on the MCAT can score as high as 250 on the exam if they dedicate themselves to acquiring a deep fund of knowledge and preparing for the exam in the *Step 1 Method* format.

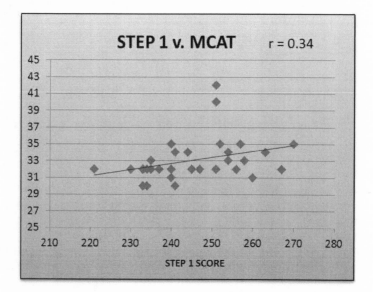

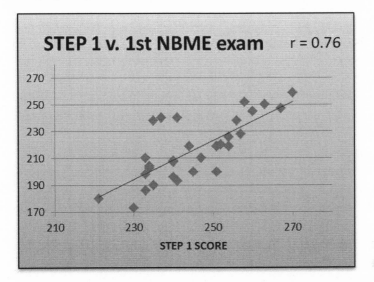

*Figure 4 Correlation Studies between the USMLE Step 1,*
*MCAT, and NBME exams*

# Guessing Effectively

The NBME prides itself on ensuring that students do not recognize everything on the USMLE Exam. Every year, they attempt to add new question formats or new question scenarios so that students are forced to keep guessing. A great example is that the NBME has a certain number of questions that are "experiment-based" questions. These are not to be confused with trial "experimental" questions that don't count toward your score. The "experiment-based" questions are routinely the toughest questions on the exam as they are not based on the classical clinical vignettes for which most students prepare. The NBME will create an experimental scenario then ask mechanistic, physiological, or pathogenesis questions requiring students to make parallels between the experiment and a known clinical entity. These questions are in addition to the typical tough three- and four-step questions about remote processes that the NBME always includes.

The fact is you will have to guess on the USMLE Step 1 Exam. Even the most prepared student will have to hypothesize occasionally on this exam. The rest of us will probably spend at least 30 percent of our time making educated guesses. The sooner you become accustomed to this fact, the sooner you will become comfortable with the uncertainty of making educated guesses. This is another major difference between "good" and "bad" test-takers. "Bad" test-takers tend not to be comfortable with guessing. This may be because they are predominantly "left-brain" thinkers and prefer to deal with facts and details. They tend to not be "creative, imaginative" types. But, if you practice something repeatedly, you will become better at it. Even the "worst" standardized test-taker can become proficient in taking the USMLE Step 1 Exam. Fortunately, the USMLE Step 1 Exam does not change its core structure from year to year. The combination of a good question bank and several NBME practice tests will create a great sense of familiarity with the exam.

# How to Cancel out Answer Choices

In order to guess effectively on the USMLE Step 1 Exam, we must do it in an educated and step-wise fashion. If you practice this style of deduction, it will become second nature. If you cannot arrive at the answer independently, your main goal will be to cancel out wrong answer choices. After canceling out choices that have the lowest likelihood of being right, you can choose one of the remaining choices that appears to be the best choice. Choosing among two to three answer choices increases your odds of being correct from 20 percent to between 33 and 50 percent. The following paragraphs focus on strategies regarding how to cancel out answer choices.

First, understand that the Step 1 is divided into different disciplines. Although questions may commonly touch on various topics (e.g. biochemistry and pathology), the question is usually rooted in a particular discipline. Then, understand each student will have their own strengths and weaknesses. Students should be able to make bolder and more confident assumptions in areas of strength than in areas of weakness. For example, if I am strong in pathology but weak in biochemistry, I will approach guessing in the two subjects very differently. If I am looking at a pathology question's set of answer choices, and I've never seen answer choice C before, I can confidently cancel it out because it is likely an element on the fringe or a topic in a different discipline. On the other hand, if I see answer choices I've never seen before in a biochemistry question, these could likely be the answer as I'm much weaker in biochem. The strategy dictates that these topics could be fairly commonplace in biochemistry, but given my relative inexperience in the subject, they have not crossed my radar. If you are a student who has relative strength in most topics, you can apply this strategy across most topics.

As a rule, focus on what you know about the topic. To what system/subject does the topic belong? If it's a question about the cardiovascular system, focus on what you know about cardiovascular pathophysiology. Think about the common themes, mechanisms, and pathogeneses that exist in cardiovascular pathology. For example, you could have a question asking about the mechanism of pulmonary edema and effusion that occurs in congestive heart failure. You may not know the answer to this. But you may remember how peripheral edema works and that an imbalance of Starling forces such as hydrostatic and oncotic pressure are to blame. If you see this as an answer choice, it is a definite possibility and should remain as an option.

This example illustrates that you will have to rely on your instincts and hunches a great deal when guessing on this exam. You will have covered thousands of topics and details by the time you sit for the Step 1. Much of the material you have covered will not be at the forefront of your mind. If someone asks you about a particular topic, or if you get a question on it, you may not be able to explain concepts in detail. You will, however, have a vague familiarity with the concept and will be able to comment on associations the topic may have. The way that we store memories often is in association or context with another element. For example, although you may not be able to describe characteristics of Vibrio vulnificus, you vaguely remember that it is associated with gas gangrene. Oftentimes, this is enough to help you arrive at an answer. In the principle of "Free Association," we teach you to follow up these hunches and vague recollections to complete the framework during your study period. You would then find out that VV is a gram (-) bacteria related to cholera that is associated with contact with shellfish and aquatic environments. You have to be confident and rely on this unconscious/implicit memory.

You can also back into an answer in an unorthodox way by knowing more facts about a particular topic than are shared in the question. For example, a common question describes a patient with a history of BPH who is treated with nasal decongestants and now presents with difficulty urinating. The NBME asks: what receptors are the cause of the patient's symptoms? Some students know exactly which receptors are responsible for the urinary hesitancy and can describe the pathogenesis of the problem. The rest of us will try to use what information we know to get at the answer. You may know that a common medicine used to treat BPH (benign prostatic hypertrophy) is prazosin. Prazosin belongs to the class of medicines known as alpha-blockers. The "alpha" in alpha-blockers refers to alpha-adrenergic receptors. Thus, it would be a safe assumption to pick alpha-receptors given that they are usually involved in patients with BPH.

Again, if you were to encounter this example during your study period, you would work out the framework. You would find that the prostatic urethra of the bladder and internal sphincter of the anus are innervated by the autonomic nervous system and contain alpha receptors. In the fight-or-flight response, a surge of norepinephrine and epinephrine floods the body, activates these smooth muscle receptors, and triggers a contraction so that you don't urinate or defecate on yourself. A common ingredient in nasal decongestants is phenylephrine, which is an alpha agonist. As a decongestant, it causes vasoconstriction of blood vessels in the nasal mucosa, thus decreasing edema and congestion. As a side effect, it can travel in the blood stream to the prostatic urethra and stimulate contraction, thus causing urinary hesitancy as a side effect. Conversely, in BPH, the prostate is enlarged and already causes an obstruction to urinary outflow from the bladder. To help counteract this, we give BPH patients alpha-blockers so that their prostatic urethra remains relaxed, and the enlarged prostate is the only obstruction to urinary outflow. As you may have guessed, another side effect of these alpha-agonists and antagonists is their activity on blood vessels. They can cause vasoconstriction and dilation of blood vessels leading to hypertension and hypotension respectively.

As you can see from this example, you do not need to know the entire framework for the topic in order to arrive at the correct answer. That said, your goal should never be to cut corners ahead of the test. The use of question strategy is meant to be used only during the exam once you have prepared to the best of your ability. Your ability to use question strategy will be much more robust if you have a large fund of knowledge on which to draw. Simple associations can be very powerful in canceling out answer choices and arriving at answers in unorthodox ways.

# Changing Answer Choices

One of the most common and classic questions we get is when should students change their answer choices from their first guess. You will hear students lament that they are racked with indecisiveness when it comes to choosing between two answer choices. They will switch their choice back and forth, an exercise wasting one to two minutes as they debate the merits of each answer. Further, their indecision will be carried on with them to the next question, thus poisoning the thought process as they continue to debate the original question.

There has been some research done in the area of changing answer choices, and it is validated by our empirical observation. A central assumption about this phenomenon is that your hunch or first guess is rooted in your implicit or unconscious memory that recalls a vague association. Often this hunch is rooted in fact that you had been exposed to long ago, but the concept never migrated to your long-term memory. This hunch is valuable, but it is not as valuable as established facts that reside in your conscious or explicit memory. Thus, you should change your answer choice if your decision to change your answer is based on new, remembered, factual information.[6]

There are two general scenarios during which a student will change their original answer choice. The first scenario is when the student recalls a new fact. Based on this newly remembered aspect or association, they now believe a second answer choice to be a better choice. Presumably, the remembered fact arising from the

---

[6]    Benjamin, L. T. et al. Staying with the initial answers on objective tests: Is it a myth? Teaching of Psychology. 1984: 11, 133-14.

student's explicit memory is credible and thus should be utilized. The student can confidently change their answer to the better choice. The second scenario is when sheer indecisiveness causes the student to change their answer. They will look at all the other answer choices and assume that other answer choices "look" better. This assertion is not based on new information, but rather on the fact that the student is not confident with their hunch and first guess. We've addressed the concept about why better "looking" answers tend to be crafted by the NBME to be low-hanging fruit used to distract students. The NBME's correct answers are all selected through a step-wise, fact-derived process. You will rarely obtain a correct answer by picking the "attractive" choice.

The skipping strategy can sometimes aid in remembering remote facts for a particular question. Often, after skipping a question, you may come to a question that actually retains useful information or that may remind you of an important fact pertinent to a skipped question. By practicing the skipping strategy, you will have time to return to this question and apply the information and change your answer choice.

As always, visit our online course to view the most recent modules for perfecting your question strategy. Question strategy can be a very powerful skill if used appropriately. Make sure that you do your best to prepare prior to going into the test. But when you are in the test, act swiftly, decisively, and confidently.

# Section 3

# Study Strategy

# Chapter 7

## How to Study for the USMLE Step 1 Exam

### What and How to study?

This is the portion of the Method where you will spend a predominant amount of your time. In this section, we cover which resources you should use and how you should use them. Your goal is to acquire a deep fund of knowledge of the basic sciences and to retain this information. You should be able to apply this information in the form of a multi-level or abstract question when queried by the NBME. Your goal will be to associate information across disciplines and topics while anticipating themes that the NBME will likely test. You want to spend a majority of your time in activities that will prompt you to actively engage and minimize the time spent in activities where you passively consume information.

### The Central Text

There are countless content-based sources that you can use. They tend to be divided into two categories: comprehensive texts (e.g. *First Aid*, Step-Up, and Med Essentials) and specific, subject-based texts (e.g. the BRS, High-Yield, and *Rapid Review* series). As a rule, your study should be centralized around one single, comprehensive text. This will be the text that you will constantly refer to throughout your study period. The text should be a substantial size but should not be too large that you cannot read through it all two to three times. The text should be comprehensive and succinct enough to review the major topics covered on the USMLE Step 1 Exam.

# Amending the Central Text

You should not simply read the central text; you should amend it. Because of their need to cover all topics, the central texts cannot go into the necessary detail about each individual topic. Further, these texts are not able to cover every topic within the scope of the exam. You must amend the text to include the vital frameworks. Your goal is to convert your central text into a central bank of facts. You want to be able to deposit and withdraw these facts at will. You will encounter thousands of facts through questions and other sources. Unfortunately, you cannot retain all of them. You cannot rely on your mind to recount all of these elements which are sources for potential questions.

You may have the tendency to write every fact/statement you encounter into the central text. Understandably, you cannot do this for fear of creating a mass of useless knowledge. Remember, what you are doing is adding essential facts to a database of existing facts; therefore, there are only so many more essential facts that were left out by the central text. Specifically, you should be inserting missing framework elements for existing topics and one-liners for topics not included in the central text. As a general rule, the facts that come from question-bank questions should be amended to your central text. You should hesitate to write down facts acquired from other texts unless you feel it is absolutely necessary. The limited space in the margins and the back of the central text is a good restriction. Some students add blank pages to their central text. This is okay, but stop short of creating another text or using other notebooks. You want to make sure all of the information you are responsible for remains in one text.

Throughout your Board study, you will have vague recollections of concepts and past facts when you encounter new material. This is because medicine is full of patterns and mechanistic paradigms that overlap. It is extremely important to not let those vague recollections remain vague; you must enforce them and cover the frameworks for these concepts. Integration of various topics is crucial when studying for the USMLE Step 1 Exam. These unifying themes are the fodder for NBME-crafted questions. Refer to your central text, review the framework, practice Free Association, and find the integration points with associated topics. You will not have reviewed only the vague fact; you will also have reviewed an associated topic. If you practice this habit—always placing key facts in the appropriate area of your central text—you will increase your retention immensely. Learning is about repetition and association. As you review your central text with the amended facts over and over, your fund of knowledge will slowly reach the mastery level necessary to score high on the USMLE Step 1 Exam.

# Seem like a Lot of Repetition?

Remember, the purpose of the *Step 1 Method* is to extract extraordinary results from prior ordinary performance. Many students who have used the Method to reach success had done the minimum in the first two years of medical school to get by. When they encounter the Method, some are taken aback by the amount of detail and repetition that exists in its steps. They realize that studying in this way requires more time than they would have otherwise spent. What they also realize is that before utilizing the Method they were average students, but after using the Method they become above-average students. In only a few short months, they are able to catapult themselves into the upper percentiles of their class and the nation. It is important to understand that as you learn the techniques in the Method, you are relearning how to learn. After several days of using the Method, you will begin to see the repetitive patterns forming. You will slowly learn that medicine is a finite set of facts and should not be feared. You will slowly gain confidence that you will be able to do exceedingly well on the USMLE Step 1 Exam. But in order to reap the rewards of a great score, you must put in the time and effort of preparation.

We recommend that throughout your Board study you review your central text at least two to three times. In addition, in the days before the exam, you will embark upon the Comprehensive Review Period. During the CRP, you will scan over your central text and review all the notes and amended facts to jog your memory for questions you have seen in the past.

# Which Central Text to Use?

We currently use *First Aid for the Boards* as the template for the central text. This book provides a nice overview of all subjects without going into painstaking detail. Anecdotally, across the nation, students will tell each other that all you have to do is "memorize *First Aid*" for the Boards and that is adequate preparation for the exam. Unfortunately, that logic is faulty in two areas. First, because *First Aid* is a comprehensive text, it cannot go into the necessary detail in all areas to cover the potential questions that the NBME writes. The authors themselves acknowledge this and advise readers accordingly. Secondly, the simple task of memorizing all 500+ pages of the *First Aid* book is nearly impossible. In addition, your time will not be well spent memorizing a book that will not be on the test. The test is made of three-step and abstract questions crafted from medical knowledge. Data must be taken from the central text and presented in a question in order for it to be retained effectively. This is the only way you will improve your performance on the exam.

Later in this section, we show you how to use your central text while working questions.

# Supplemental Sources

While the central text will be the basis of your study, supplemental sources that go into more detail are necessary to fill in the conceptual gaps between ideas. Oftentimes, the central text will just present a list of symptoms or treatments without explaining why these occur or are effective. It is sometimes necessary to have a more detailed source to scan in order to understand concepts that cannot be explained with bullet points.

Many students have a large collection of supplemental sources—one for each subject acquired through the first two years. The mistake that many make is feeling obligated to read through all of these during their Intensive Study Period. Worse, they tell themselves they won't begin questions until they finish reading the supplemental source. Realistically, you will not have enough time to read through each supplemental source and still have adequate time to work enough questions the right way to reach mastery level.

You should spend adequate time before beginning your Intensive Study Period thinking about how and when you will use your supplemental sources. We want you to take inventory of yourself and gauge your strengths and weaknesses. As a rule, you should prioritize use of supplemental sources in areas of weakness. In areas of weakness, the extra context will be useful to explain concepts that you cannot readily grasp by working questions and using the central text alone. On average, students will use between two and four supplemental sources in total. If you have the time to use more than this number, take care that you are not stealing time from doing questions in the right way.

When selecting a supplemental source, make sure the information is easily accessible through a detailed table of contents and that it has an index for rapid access to facts. This text should go into adequate detail but not reach textbook-level denotation. Instead of reading the supplemental source line for line, you should attempt to scan the text. By scanning, we mean that you should glance over the topics located on each page and determine if you are familiar with them. If you are not, check to see if those facts/diseases/syndromes are in your central text. If they are, make sure that the central text covers the framework for that topic. If the topics/facts are not in your central text then they are less likely to be high-yield and therefore less important to know. The important thing is that you know these topics exist and are able to place them in one category such as metabolic diseases or cardiovascular diseases. For example, simply knowing that Hurler's syndrome is a lysosomal storage disease and McArdle's disease is a glycogen storage disease will take you a very long way: in question strategy, you will be able to cancel out wrong answer choices because you will know which classes they fall into. Thus, it is not important to know every single detail about a topic, especially when the topics are on the fringe of core USMLE Step 1 topics.

Do not use textbooks to study for the Board exam. The unnecessary detail will bog you down and limit the amount of time that you have to work questions the right way. If cost is an issue, most medical schools have many supplemental sources available that you can borrow from the library. Do not worry about using review texts that are one to three years out of date; not much has changed in the world of testable medical knowledge.

# Studying with Questions

The *Step 1 Method* is a question-based study method; the bedrock of your preparation will be based on working questions the right way. In our seventh year of existence, I have learned to be very careful about how I talk about working questions. In the past, students spent a majority of their time reading supplementary sources and a minority of time working questions. In 2005, our goal was to get students to start spending a majority of their time working questions. In 2013, we've seen a huge shift in the way that students study and some startling findings have come to light. Working questions has now become the foundation of most students' study patterns. Surprisingly, we're finding students working through a number of question banks and still not reaching optimal practice scores. It became very clear that in order to achieve optimal results, you must work questions the right way. It is easy to work through a question, but it is much harder to learn and retain the necessary information from a question. We find that many students do not have the patience to go through questions in a step-wise fashion.

A fundamental mistake many students make is using their question banks as an evaluation tool to test their knowledge or speed. When students test themselves before they are ready, they forfeit the opportunity to use the question banks for the immense amount of knowledge they contain. The *Step 1 Method* Study Strategy focuses on gaining knowledge from questions and maximizing retention by reading after working through questions. Furthermore, the majority of your reading of content-based sources will be driven by the actual questions you complete. The majority of your time should be spent working actual questions or reading after questions.

# Why Are Questions So Important?

The theory behind this technique is that learning information presented in a question involves active concentration, emotion, and contemplation. When a fact is learned in this manner, you are much more likely to retain it—in contrast to trying to remember one of a hundred facts read on one of hundreds of lines, on hundreds of pages. Tying information to emotional stimuli greatly enhances the retention of material. Imagine how enthused you are when you either get a question right or wrong. When you get a question wrong, part of you is mad or frustrated, but there is also an urge to find the right answer so that you never get that question wrong again. This energy and motivation are invaluable tools to use because studying for an exam tires out many people before they are able to reach their true potential. Harness this energy; seize the opportunity to improve and stay motivated. Never get discouraged if you are getting questions wrong. Instead, understand that you are actively finding holes in your knowledge base. With every question you get wrong, you are learning a new fact, and in essence you are raising your potential score. If you look at it this way, then it makes perfect sense to work as many questions as you possibly can before the test. If you do this, you will be able to expose yourself to a maximum number of holes in your knowledge.

A main mental block that medical students have with the USMLE Step 1 Exam and medicine in general is they are intimidated by the sheer amount of knowledge covered. When viewing the field as a whole, it can appear infinite and unconquerable. This feeling of inadequacy forces many students to give up and throw in the towel. On the contrary, the high-yield information covered on the USMLE Step 1 Exam is very finite; with adequate preparation, you can learn exactly what you need to in order to do well. Your preparation will build on the curriculum you learned during the first one-and-a-half years in medical school. We advocate that you start studying six months prior to your test date, while you are still in school, in order to feel comfortable with the material prior to beginning your Intensive Study Period. In this way, you will have seen everything at least once before reviewing key, high-yield areas that are most likely to show up on your test.

# The *Step 1 Method* Format of Working Questions

Most students who study for the USMLE Step 1 Exam do work through questions; unfortunately, they are doing them wrong. One of the central tenets of the *Step 1 Method* is that you must relearn how to work questions. Oftentimes, students fly through a block of 46 questions and see how many questions they get correct. When they see their percentage correct, they will either get happy or sad. Unfortunately, you do not learn any new contextual information from this percentage. If these students are diligent, then they will go back over their questions. But which questions will they look at? They will look at the questions they got wrong. Newsflash: sometimes we get questions right for the wrong reasons. These students will miss major areas of hidden weakness by not reviewing every question. Further, after 46 questions and over an hour later, you likely forget the thought process you went through to get to that particular answer. Thus, you are no longer emotionally or conceptually focused on that question and the subject matter it represents. Finally, the temporal aspect of doing 46 questions tends to take effect and students just want to be done; they will often cut corners in their learning just to finish a block of questions.

To correct for these shortcomings, we've developed the *Step 1 Method* Format of Working questions to maximize retention and performance:

1.  Do questions one at a time, in tutor mode. After attempting the question, immediately read the answer explanation.

2.  Go to the corresponding section in your central text, and read the subject heading dealing with the topic tested in the question. (e.g. If the question is about coronary artery disease, read the section about coronary artery disease.) Is the key framework info contained in the central text? If not, annotate the missing information concisely in the margin next to the topic in the central text. Use short, incomplete sentences of six to seven

word associations. Do not copy down the answer explanation. Do not write paragraphs. Keep your annotations limited to framework information about the topic in the question.

3.  Look at the wrong answer choices. Do you know what classes they belong to? Read those explanations as well. Feel free to review those areas in your central text as well, but do not annotate on wrong answers as this will take a very long time. If these topics are important, there will be questions on them. If you feel there is a key piece of framework info left out of the central text, it is okay to annotate briefly on these topics.

4.  After completing these steps, make sure you are confident about the framework info for that topic. Imagine what other elements of the framework they could have tested you on. Once you have done this, you are ready to move on to the next question.

5.  Never worry about speed. Always focus on learning the material and retaining the important facts. The best way to finish a test quickly is to know all the answers.

6.  It should take you between 5 and 15 minutes to work each question in this fashion. If you find a glaring area of weakness that requires a lot of review in the central text and supplementary texts, it is okay to spend as much as 20 minutes on a particular question and topic. Your timing will speed up on subsequent questions on the same topic. For example, it will take you 15 minutes to do your first question on cystic fibrosis as you will need to build out the framework. The following questions on cystic fibrosis won't take nearly as long because most of the annotation will be completed.

7.  Be mindful of esoteric topics that you have never heard of and that are not in the central text. Keep your review of these topics limited to building and annotating the framework for this topic. The remote topics tend not to be high yield. Students tend to spend too much time on these topics and not enough time on their major weaknesses, which are often covered.

# How to Work Blocks of Questions

You should work questions in blocks of 46 as they correspond with the topic group you are covering. If you're working questions while in school, they will correspond with your school curriculum topics. If you're in your ISP, the questions will correspond to the subject/system you are covering at that point in time. You should not randomly work questions as many students try to do, attempting to simulate the test environment of jumping from topic to topic. Again, they are mistakenly attempting to use the question bank as an evaluation tool instead of an educational tool. You don't want to evaluate yourself before you have learned all the material. The best way to finish a test on time and score a high percentage is to know the answers to the questions. By testing yourself before you have learned the material, you can assure yourself low marks on both measures. You will get plenty of exam simulation with the three or more NBME practice tests that you take throughout your ISP. At that point, you will have had enough time to accurately assess your competency in all topics.

After finishing the number of questions within a topic group, there will be a significant number that you have gotten incorrect. We recommend reworking these incorrect questions before moving on to the next topic group because the timing of reworking your incorrect questions may be an issue. For example, if your school curriculum jumps to a new topic group before you are able to rework your incorrect questions, you should appropriately follow the curriculum. Understand though that you will have a question debt to repay. Your goal will be to rework these incorrect questions along the way prior to resetting your question bank. These incorrect questions are the largest holes in your knowledge base; you want to make sure that you master them before moving on.

In the *Step 1 Method*, we advocate that students should attempt to get through their question bank twice. This will enable you to see most questions twice, and some questions three and four times. There are over 2,000 questions in the leading question banks. You will be working these questions over a three- to six-month period. It is unrealistic to assume that you will remember the topic of question #2 by the time you get to question #2,100.

Further, the goal, as we've stressed regarding working questions, is not to memorize the questions and answers. Your goal is to learn the frameworks behind the topics. Thus, when you are going through the Qbank a second time, your purpose will be to see if you have mastered individual topic frameworks. It's important to understand that you will not see the same Qbank questions on your USMLE Step 1 Exam. You can be assured that you will see the same topics tested on the exam, but the chances are that the NBME will choose a different element of the topic framework to test.

In our Preparation Strategy section, we take a closer look at "When to Reset the Question Bank," a dilemma in which many students find themselves.

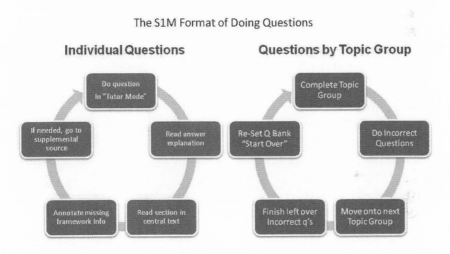

*Figure 5 The Step 1 Method Format of Doing Questions*

# Which Qbank to Use?

There is much dispute in the marketplace over which question bank you should use. Currently, there is little doubt that the two most established products are:

1.  USMLE World
2.  Kaplan Qbank.

Both are effective banks and will supply you with the necessary information. Based on student feedback over the past five to six years, I would rate USMLE World as the better Qbank based on the quality of the answer explanations and the similarity of questions to the actual exam. But the way you should work the question bank does not change. Your use of the question bank should begin immediately, and you should finish the question bank well before you are ready to take the exam. Even further, we recommend that you aim to get through your question bank twice in the *Step 1 Method* format. Please note that all Qbanks are not created equal, and any one Qbank will not do. We have purposefully left out other Qbanks that we could have mentioned. You can take this as a sign that we do not recommend them. If you want complete coverage of the Qbank offerings, go to our website and check out our "Content Corner." There, we examine and provide reviews and student feedback for many different forms of prep resources.

# One Qbank or Two Different Qbanks?

In order to reach your potential and achieve Boards mastery level, it is important to go through your Qbank a second time the proper way. Too often, you will hear of students starting the question bank in the month before the exam. They either rush through it or can't finish it all, leaving huge holes in their knowledge base. You cannot afford to waste the most valuable resource for your test preparation. Fortunately, the major Qbank producers provide six-month subscriptions to their test banks. These are well worth the cost and should be the major purchase of your test prep materials.

Unfortunately, many students spend thousands of dollars on overpriced packages which include online, content-based lecture series and unnecessary books. These resources require too much time to cover, and they do not lead to direct retention of the material. The answer explanations located in the question banks do an excellent job of explaining answer choices and important concepts. Furthermore, if you have trouble understanding a concept or need more detail, refer to a supplemental source for that subject. If that does not suffice, Google your topic and utilize credible, online sources such as e-medicine for rapid access to facts. This is much quicker than watching a lecture online and hoping the lecturer covers your desired topic in adequate detail.

A common question I receive is from students who are diligent enough to finish their Qbanks with time to spare and wonder if they should purchase another Qbank. Put simply, your goal is to pick one of the approved Qbanks and master it in the *Step 1 Method* format. Many students are overzealous and purchase multiple subscriptions to different Qbanks simultaneously. I have to shake my head when I hear about students switching back and forth between question banks. They end up with two half-finished Qbanks and usually do not have a regimented format of recording the information they have learned.

Another common scenario I hear about is students being advised (usually by upperclassmen) to purchase the Kaplan Qbank (or other Qbank) during the regular school year, then purchasing the USMLE World subscription during their ISP. The problem here is that students will not remember all of the key topics after going through a Qbank once. True mastery requires going through a Qbank twice, the right way. Why waste half of your time with an inferior question bank when you run the risk of not being able to fully utilize the best on the market? By waiting until the last one to two months, you run the very real risk of not being able to complete the Qbank and will certainly not have enough time to repeat it. Another real risk is that when students move to a second Qbank, they may never get adequate review of their weaknesses. These students will fly through the first Qbank and then enter the second Qbank at a hurried pace. They may make it through a significant portion of the second Qbank, but they may only review strengths and never hit their weaknesses hard. Going twice through the same Qbank in the *Step 1 Method* format ensures you will hit all of your weak areas. In your second pass through the Qbank, your primary concern is to review frameworks and to dive deeply into weaknesses to confirm mastery of these high-yield topics.

If you are extremely ambitious and prepared, you will complete your question bank twice with time to spare. Then you can contemplate purchasing a short subscription to the other major question bank product and begin doing that one. Working through a Qbank three times is overkill, as you will remember all the questions. The other major Qbank product will have a different style and focus. The topics are usually the same, so more practice is good.

# Attacking Weaknesses

A central tenet of the *Step 1 Method* is learning how to diagnose and attack your weaknesses. To reach your potential on the USMLE Step 1 Exam, you must become self-directed and introspective. You must be your own coach and director. You must be honest with yourself and diagnose your strengths and weaknesses on a daily basis. Don't rush through topics and questions. When you sense you don't know something as well as you should, stop and learn it. You must learn—if you haven't already—how to teach yourself important facts. Given the enormity of information on the USMLE Step 1, a practiced skill is discerning which facts are important and which aren't. Fortunately, your Qbank and the *Step 1 Method* Framework aid you in this task. If the question bank has a question on it, then it is a high-yield topic that you should review and master. You will see topics that are either not covered in the *First Aid* or the Qbank, or both. The likelihood that these topics are high yield and well represented on your Step 1 Exam is rare. Thus, you should spend a maximum amount of time reviewing and mastering the topics and concepts that you encounter among the 2,000+ questions in the question bank and those covered in your central text.

# Getting Questions Wrong Is Good?

When you begin to work through questions early, you have the opportunity to find your weaknesses and attack them with persistence. In turn, because we are using the question bank as an educational tool rather than a measurement tool, you should not worry about your percent correct. Oftentimes, students will get discouraged about scoring below 50 percent on some question blocks. They quickly become frustrated and begin to dread studying and working Qbank questions. Then they vow not to work questions until they "read" first. They postpone reading through the *First Aid* section or central text for one to two months and never start working through questions. The next thing they know, they've only finished 10–20 percent of the Qbank and their Intensive Study Period is less than a month away. This is the typical situation in which most students find themselves.

In the *Step 1 Method*, you will be learning about a question when you get it wrong. Then you will rework that incorrect question, ensuring that you learn the topic. Then, hopefully you will reset your question bank and work that question a third time. You can see that through this process, you will have reviewed that topic framework three times. This sort of repetition gives students confidence. You will hope the NBME tests you on that particular topic. Many students who have scored 250s on the USMLE Step 1 Exam received mid-50 percent scores on the Qbank the first time through. I've often encountered students who score 70–80 percent on the Qbank the first time and thus coast into the exam confidently. They are disappointed when they barely score above average on the actual exam. This example shows why many of our former students proclaim that the percent correct on the Qbank the first time through means nothing.

# The Principle of Free Association

Another skill you must master in the process of doing questions is the Principle of Free Association. The Principle of Free Association is based on the premise that most topics belong to a larger class of topics that all share similar characteristics. The frameworks for these topics only differ in a few elements. As you can see, this is a great opportunity for the NBME to test the depth of a student's knowledge.

For example, what's the difference between dermatomyositis and polymyositis? Sjogren's syndrome and mixed connective tissue disease? Limited and diffuse scleroderma? You should be able to compare these across the *Step 1 Method* Framework and point out the similarities and differences. You would have learned these subtle differences and similarities if you had done Free Association for connective tissue disorders. When the NBME examiners want to test the depth of your knowledge, they will make these similar topic elements the wrong answer choices. Students find themselves stuck among two to three answer choices because their knowledge base—usually rooted in memorization—does not go far enough.

In order to master these gray areas, you should not only learn about topics in isolation, you should also attempt to associate each topic with similar topics and be able to compare their similarities and differences at will. This is the Principle of Free Association. If you get very good at this, the amount of Free Association that you can do is unlimited. You can compare bacteria which have similar toxin mechanisms, then those that have capsules, then those that have vaccines associated with them, then those that cause the same disease processes, etc. Very quickly, you will find that there is an association that you know exists, but you can't make the connection. That is a hole in your knowledge base. Many students will feel rushed and shrug off that hole to work through more questions. We want you to stop and make that association.

The facts tested by the NBME examiners rarely change, but the way they ask the question and the element of the framework they test will change from year to year. If you understand the framework behind each topic and question, you will be able to answer any phrasing or situation as long as you come back to the central idea.

Once you become familiar with the medical pearls and central ideas, you will be able to anticipate the possible questions the examiner can ask. For example, if there is a question about the beta-blocker class of medications, it will usually deal with one aspect—let's say side effects. But as you learn about beta-blockers, you will also learn indications for use, mechanism of action, and pharmacokinetics. So when you see metoprolol in the question stem, you can start to imagine on which area the NBME will focus and go over those particular areas in your mind. If you are weak in a particular area, review that area because it will become a question in the future.

# Chapter 8

## Studying during the Second Year of Med School

### Study for School Curriculum, the Boards, or Both?

We advise that students begin studying for the USMLE Step 1 Exam five to six months out from their test date; this is a major methodology departure from what is recommended by most medical schools in the United States. Many school administrations tell their students that they shouldn't start studying for the Boards until the end of their second year, based on the claim that a school's curriculum alone is the best preparation for the USMLE Step 1 Exam. This advice is based largely on the administration's belief that their curriculum is truly adequate and the desire for their students to pay attention to the school's curriculum. All too often, students disregard their school's curriculum and focus only on USMLE Step 1 Exam preparation.

When working with our school clients, I often find myself in the middle of this debate. Every school's curriculum is going to be slightly different, but all schools are accredited by the same organization and therefore will largely cover the same material. The major difference I see among schools is how students are examined. Some schools will have multiple-choice exams that attempt to mirror USMLE-style questions. Other schools will have essay-based exams that look nothing like the USMLE Step 1 Exam. Each curriculum has its champions and detractors. In all cases, I know that individual administrations feel that they are giving their students the best chance for success. But in both cases, because of the sheer enormity of information out there, students must begin studying for the Boards several months prior to their Intensive Study Period. The goal will be to spend a majority of your Intensive Study Period reviewing concepts and attacking weaknesses instead of learning new topics. It is inevitable that you will have to learn new things during your ISP, but the more time you have for review, the better chance you have of reaching mastery level.

We believe that you should fully engage with and participate in your second-year coursework. Ironically, students who ignore their second-year curriculum to focus on the Boards are often ineffective. I believe most medical students do better with structure. By removing all structure, these self-guided students tend to underperform in both their curriculum and their Board study. In the *Step 1 Method*, we create a plan to start working questions along with your coursework. This way, your Board study is adding to your school curriculum work and vice-versa.

For two major reasons, we do not recommend starting to work questions until you are roughly six months out from your test date. First, the second year of medical school is very difficult. It will take you several months to gauge the amount of work required to learn the information and do well in your classes. We believe that your curriculum should take priority to Board study. Second, the more time you have, the more time you have to forget information you have learned. Simply put, you will likely forget information if it has been nine months since you first learned it. You will see the same phenomenon in your Intensive Study Period. For most students, we do not recommend studying for longer than five to six weeks. As you are rapidly learning information, by week six, you begin to forget topics you learned in week one.

Students should not rest assured that their curriculum is total preparation for the USMLE Step 1 Exam. While doing their Qbanks, students commonly encounter topics they have never seen before. Two years of curriculum is a limited time, and faculty must pick and choose which topics to include and exclude in their curriculum. Students then have to take it upon themselves to teach themselves this information; this can be a startling situation. It is for this situation that we created the *Step 1 Method* Framework: so that students would understand what they need to know about each topic.

To make matters worse, many schools are limiting students to only three weeks to study for the exam. In addition, this time often overlaps with vacation time; therefore students are forced to choose between vacation and dedicated Board study time. Until the students' performance on the Boards is held in high regard, professors will teach what they believe is important, not what the NBME thinks is important. Therefore, you must take your performance on the Boards into your own hands.

# Working Questions during School

As mentioned, many school administrators will discourage students from starting Board study early for fear it distracts students from their curriculum. If done correctly, the opposite is true. There is a way to improve not only your performance on the USMLE Step 1 Exam but also your current coursework. The enormity of the challenge ahead of you and the variability in the amount of time schools give their students require that you begin your Board study during your school year. For most students this simply entails reviewing a central text, such as *First Aid*. We recommend that you begin to work questions six months out from your test date, as they correspond with the curriculum you are covering. Most topics covered by the question banks are also covered by your professors in school. The major difference will be in the level of detail each forum will reach. Your professors may go into painstaking detail in some topics, while you will get only one to two questions about the topic and limited coverage in the central text.

If you start to work through questions early, you will quickly see exactly how much of the information on your course exams is actually represented on the USMLE Boards. This can either be gratifying or frustrating, depending on the style of the professor. For many students, it is a battle because their professors lecture on esoteric topics and their exams focus on remote facts that are not considered high yield by the boards. Understandably, the NBME examiners will focus their questions on high-yield, important concepts because they must cover two years of subject topics in one exam. In this fashion, test-preparatory products such as *First Aid*, USMLE World, and Kaplan Qbank focus on areas that are high yield and most likely to show up on your test.

Starting out, it will take you awhile to get through a block of questions. You should aim to do one block of 46 questions per week. For example, if you are covering cardiovascular pathology that week in school, you should select 46-question blocks from this topic group until you finish it. When choosing special topic groups, the Qbank will tell you exactly how many questions are contained in those categories. Practically speaking, you should expect to work through two to three questions during the weekdays, then 15–20 questions on each weekend. Class days will be invariably busy, so adding a few questions during lunch or prior to bed is good practice. You will largely be doing most of your question work during the weekends. On a week-to-week basis, don't get frustrated if you fall off schedule. Try and end the month with the necessary number of blocks completed.

After about two months, you should pick up your question volume to two to three blocks of questions per week. Your speed will increase, and this will be an achievable goal. As your second year ends, your professors may choose to provide NBME shelf exams as your final exams. There is no better way to study for these exams than to work through questions in the *Step 1 Method* format. Your goal will be to finish at least 50 percent of the Qbank by the end of the second year. Many students finish between 70 and 80 percent of the Qbank, while others finish the entire Qbank prior to their ISP. Everyone has a different capacity for how many questions they will do. If you are able to finish at least 50 percent of the Qbank, you will be able to complete the remaining portion during your ISP and start the process of reviewing.

Understandably, to complete all these questions, you will have to be working harder than you did in the first half of the second year. Fortunately, you will be much smarter than you were in the beginning of the year and will be able to cope. Remember, you are focused on a goal to reach your potential on this exam. As always, never speed through the questions; focus on learning the material and being able to visualize and anticipate the next question on a different framework element of that topic. By attempting to read the NBME's examiners' minds, you will start to see potential questions wherever you look and will train yourself to anticipate question stems when reading particular topics. Once you start doing this, you will undoubtedly begin to see your predictions come to life in real questions.

# Chapter 9

## Other Supplemental Sources

For a complete review of the most commonly used supplementary multi-media resources, visit our website's "Content Corner." There, we provide reviews, student feedback, and links to learn more about the resources. In the sections that follow, we highlight the most asked questions about supplementary sources other than books.

## Live Lecture Courses

One of the oldest forms of USMLE Step 1 Exam preparation is the live lecture course. These courses are traditionally marketed to international medical graduates and US medical graduates who either struggle with their curriculum or fail the exam. These courses tend to last for three weeks and are a series of daily six- to eight-hour days of lecture after lecture. The preparation company hires a staff of instructors that usually have faculty positions at various medical colleges across the country. Each instructor will have a subject of expertise and can become quite famous in these circles. The most famous is undoubtedly Dr. Edward Goljan, a pathology professor from the Oklahoma State University Center for Health Sciences.[7] He initially taught for the Kaplan prep course then transitioned to the Falcon Prep Course.[8] He authors the *Rapid Review* of pathology texts. He is most known for an unauthorized recording of his pathology lecture series while working for Kaplan.[9]

---

[7]   http://elsevierauthors.com/edwardgoljan/
[8]   http://www.falconreviews.com/v3/live-instructors.html
[9]   http://en.wikipedia.org/wiki/Edward_Goljan#cite_note-Falcon-2

These live courses can cost anywhere between $3,000 and $5,000. They usually offer lodging at the hotel site of the lectures. These packages tend to contain course textbooks and may contain a short one- to three-month long subscription to a major question bank. They may also sell supplementary materials created by affiliate instructors.

These live courses appeal to students' familiarity with review courses. Most US medical students either used a prep course to study for the MCAT or knew several people who did. In my experience, few (< 10 percent) of US, first-time test-takers attend live prep courses. On the other hand, after failing the exam, a much larger percentage (> 33 percent) of this pool of students will take a prep course. Many US medical schools will strongly suggest that students who have failed consider taking one of these expensive prep courses.

These courses are simply a high-yield review of major topics in basic science curriculum. As you can imagine, attempting to review two years' worth of science in three weeks will be largely incomplete. Students may receive unique insight from experienced instructors, but the insight will be abbreviated and will touch on only a handful of topics within a sea of information. Lectures will last throughout the morning and the afternoon. Students are often exhausted afterwards but are usually required to read along with class notes. There is little if any time for students to work through questions during this three-week prep period. Because most US medical schools give their students a maximum of six weeks to study for the USMLE Step 1 Exam, the students will return to study for about three weeks prior to taking the test. Thus, they will have only a dedicated three weeks to work questions and attack whatever weaknesses they have. They may or may not take practice exams because they are afraid of what they may find out.

Given this scenario, it is not surprising that students who take these intensive prep courses barely pass, or may fail the exam. I believe the core problem with these courses is that these courses force students to spend a majority of their valuable study time practicing low-retention activities. As has been mentioned, you retain very little by practicing a passive activity such as sitting in lecture all day. Further, US medical students have spent the last two years in medical school sitting in lecture. They do not need more of the same. True, many medical students will state that they didn't learn things well enough the first time through. Because of the speed at which topics are covered in the first two years, there wasn't enough time for facts to sink in. Well, these three-week courses are going to be even worse. You are going to fly through several high-yield topics in a matter of days. Further, you are not able to stop the instructor and ask for more clarification when you don't understand a topic.

The worst thing that can occur is that students feel overwhelmed with all of the topics and cannot slow down to further digest information. They tell themselves that they will review poorly understood topics after they complete the course. Quickly they get overwhelmed, but they have to continue to participate because they took out an emergency $5,000 loan to pay for the course. In this situation, they suffer through the remaining course. The live course then becomes a low-yield, low-retention, figurative prison of rapid review.

I believe that most students who take these review courses feel comfortable because they can read along with the notes and remember the topics covered from their school's basic science curriculum. These instructors are best at distilling complex topics into simple "need to know" facts. Students appreciate their jobs being made simple for them as they tend to prefer simple memorization of a list of facts and tables because there is a finite beginning and end of a list or table. It is much more challenging to think in an open-ended manner about medical topics. Diving deeply into weaknesses or poorly understood topics is much easier to avoid. Students rejoice when an area of weakness is not covered in the rapid review session because they no longer feel responsible for the information. They are then sorely disappointed when they see multiple questions in their Qbank about this topic.

For IMGs whose medical school curriculum was drastically different than that of the US medical schools, these courses can serve a purpose. These courses can function as an introduction to the US medical school curriculum; however, these courses alone will not be adequate preparation to do well on the exam.

## Online Lecture Courses

As the USMLE Step 1 prep market matured, it found that many students were not willing to pay over $4,000 for live lecture courses but still had the appetite for watching lectures. As the bandwidth of the Internet made it possible to deliver hour-long video lectures at minimal cost, the live lecture course was born. These online offerings come in two major varieties: the online "live" lecture series and the "high yield" series.

The online "live" lecture series is basically just a videotaped version of a live lecture. The number of hours of lectures can range from 70 to 200 hours. There may be optimized settings with a built-in slide format within the visual display, but you will see an instructor in front of a room, giving a lecture. It is roughly identical to the live lecture series without the crowded rooms, hotel booking, and communal breakfast and lunch. What it will lack though is the boot-camp mentality of hunkering down and preparing for the test. Many students prefer this solitary confinement mentality provided by live lecture courses. These online live lecture varieties will range from $1,000 to $2,500 for a one-month to three-month subscription. This is an important facet to understand. If you have only 30 days to view over 70 hours of content, you must spend a significant portion of your day viewing these lectures. So, there is pressure to rush through these lectures while you have access. Of course, these companies charge you extra to extend your subscription, leading to a very expensive and often ineffective product. These courses may or may not come with course notes and high-yield materials. They usually do not include subscriptions to the major question banks.

The same problems that occur with live lecture courses also apply to online live lecture courses. The only difference is that you are watching these lectures on your computer instead of in a hotel conference room. These courses alone do not hurt students' chances of doing well on the exam, but students often believe by watching these videos alone, they will be very prepared for the exam. They spend a majority of their time watching and re-watching videos instead of working questions the right way. The students will watch the lectures go into surface detail over most topics but won't take the time to dive deeper into their own personal weaknesses. They will pass the exam, but their score will be disappointing and undershoot their potential. Again, we have worked with several students who purchase and watch these lectures and fail the exam because of inadequate optimal preparation.

Many students ask how to use these lectures if they've already purchased them. In that case, use them as you would any other supplemental source: for extra context in areas of weakness. The lectures could potentially give you new insight into how to think about complex topics in a simple, straightforward way. Consistently though, you will find the key information covered when you work questions and review your central text. When you start to spend time using supplementary sources in areas that are not weak, you are rarely going to learn more than you would than from questions/central text and may actually hurt yourself because you will be sacrificing the time that should be spent working questions in the right way.

# Online "High-Yield" Lectures

This is a new format of online lecture course that has appeared on the market in the last two to three years. These are narrated slides that simulate a central text. Many of these offerings will actually just present slides that look exactly like *First Aid* and correspond with the chapters in *First Aid*. The narrator will attempt to explain the concepts in the text and talk through some of the graphs. They may also work out formulas and problems.

Student feedback from those who have used these resources have returned mixed results. The *First Aid* text itself is very easy to read and understand. Is it necessary to have a narrator explain it to the student? Perhaps for some complex concepts, extra explanation would be helpful, but this help arrives at the cost of listening to several unnecessary hours to find the valuable 15-minute explanation. A student is likely better off using a text supplementary source to immediately shed more light on a difficult area. Again, if these high-yield lectures have already been purchased, using them in areas of weakness would be optimal.

We've already detailed how the *First Aid* itself is incomplete in its presentation of topic frameworks: the student has to supplement the *First Aid* with their own annotations of topic frameworks from their question bank work. These high-yield lectures will also likely be inadequate over their review of all the necessary information because they are a verbal/visual representation of the central text (e.g. *First*

*Aid*). Thus, the amount of time wasted is compounded by watching lectures that are incomplete.

To be clear, the biggest problem with all forms of lecture resources is that these time-intensive activities steal time from working questions the right way. A student who invests their time and money in a lecture course is increasingly unlikely to get through a preferred Qbank twice in the *Step 1 Method* format. If a student is able to utilize a lecture course and get through the Qbank twice in the right way, then that student should have a good chance of success.

# Web Path

This is an online database of pictures and clinical cases. Using pictorial sources will help as 10–15 percent of USMLE Step 1 Exam questions will have a central photo or histology slide and will ask you questions relating to the image. Oftentimes, the NBME will just present an image, and you will have to first know what you are looking at, then answer a question relating to the disease process. If you do not recognize the image, it may be difficult to answer the question, but you should not become immobilized. Sometimes the question stem will provide enough information (i.e. symptoms/signs) for you to elucidate what the disease process is. Based on your knowledge of the disease, you can take the steps to answer the framework element being tested by the NBME. But as mentioned, why rely on chance? Do your best to review slides before coming to the exam.

# Goljan Materials and Audio Lectures

The Goljan audio lectures and high-yield notes are a "cult-classic" of USMLE Step 1 prep that many students do not know about. They are a series of 35 high-yield pathology board lectures that go over important topics and case scenarios given by Professor Goljan during a live lecture course. The lectures are distributed on the Internet in an unauthorized manner; thus we cannot endorse that you download them. That said, these lectures are often passed down from upperclassmen, so many students will come into contact with them. We will provide some guidelines on how to use these or any audio lecture series.

As is common with lecture-based resources, students gravitate toward them because they are easy-to-use, passive-learning formats. Students can just sit back and listen. You already know that this is the worst way to attack weaknesses and that it does not lead to high retention. A major mistake students make is reserving hours of their Intensive Study Period on a daily basis to sit and listen to these lectures. Students will sometimes take notes on these lectures. The precious 8–10 hours per day that you have in your ISP are too valuable to sit and listen to 15-year-old clinical vignettes. In these lectures, Goljan is notorious for listing one-liners—a list of signs and symptoms—and asking for the diagnosis. Then he may follow up by asking for another element of the topic's framework. That's it. The review is not complete, but rather a spattering of topics across all of pathology. It is mixed with his stories and humor.

If used in the right way, the audio lectures can be a nice respite from the daily grind of working questions and reading supplemental sources. The use of these audio lectures can provide an uncanny and unconscious element to your knowledge base as you learn passively. You will start to know the answers to questions because answer choices sound familiar even if you cannot remember exactly in what context Dr. Goljan mentioned them. He specializes in pathognomonic terms. These are signs/symptoms that are only associated with one particular disease process or state. Thus, when you hear these terms, you should make the immediate leap to the respective disease process.

The best way to use these lectures is during your off-time (e.g. at the gym, walking to and from school, driving, cooking, before bed, etc.). The idea being that even when you are not actively working, you will still be learning. The lectures flow at an even and steady pace and you will find it a welcome escape from the text-and-question-bank routine. Do not feel the need to take notes; just listen, learn, and watch your knowledge base expand. We recommend starting these audio lectures early so that you may be able to go through them at least two to three times. After the second time through these audio lectures, you will approach mastery level. And to further approach mastery, share what you have learned with your classmates. Once you are able to verbalize your knowledge, you will be able to own it. If you continue to passively learn without acting on and refreshing your knowledge, it will only be a matter of time before you forget those facts.

# The Internet

More and more, Internet sources are becoming a great source for obtaining reclusive knowledge. If you are not able to find a particular fact in your BRS or other supplemental source, Google it and see what comes up. The first few links will likely belong to e-medicine or other medicine websites, which are usually credible and trustworthy factual sources. If your school subscribes to *Up to date*, then you have the opportunity to use this excellent source for medical knowledge. Take care not to go too in-depth with these sources, but rather extract the necessary framework topic elements and move on. Using the Internet is much faster than pilfering through a giant textbook like Robbins.

# Section 4

## Intensive Preparation and Evaluation Strategy

# Chapter 10

## The Intensive Study Period

In this section of the *Step 1 Method*, we analyze the last weeks before the USMLE Step 1 Exam: the Intensive Study Period. Our goal in this period is to design a targeted plan to reach your potential by assessing and targeting weaknesses as you lead up to the test. First, we cover how to properly plan and execute your Intensive Study Period. Second, we explain our Practice Test Strategy. This strategy helps you determine how to accurately evaluate your performance, identify weaknesses, and target these weaknesses to reach your potential on the exam. You will also know exactly where your score stands going into the test and whether or not you are ready to sit for the exam.

## Creating an Intensive Study Period Schedule

Planning an Intensive Study Period (ISP) schedule is something that students either spend too much or too little time thinking about. You should definitely have a plan in place before you start studying, but you should not obsess over it. You should understand that the plan will have to remain fluid: you may have to adjust it as you go along. On the other hand, some students do not have a plan at all in place before they begin their ISP. These students quickly find themselves lost when they hit a road block. Without a guide to keep you on track, the slightest bit of adversity can defocus the most confident of students.

So, now that we know we must create a study schedule, where do we start? There are a number of factors you have to take into consideration when planning your study schedule. We will review them here. Please note in advance that we have created a software tool—"The *Step 1 Method* Study Schedule Wizard"—that automates the following steps. The Schedule Wizard has the ability to create a personalized ISP schedule in minutes. We still believe reading through the following considerations is a valuable experience because these steps explain how your personal schedule is created. In the event that you need to adjust your schedule, it will be based on one of the following factors.

1.    Total number of days in the study period

This will be the most important factor. You will determine how many days that you have to study and divide them up amongst the various topics. You should reserve 10–15 percent of your total time for comprehensive review at the end. You should also budget in two days for NBME practice tests. An optimal amount of time for your Intensive Study Period is four to five weeks—longer than that and you run the risk of forgetting facts covered at the beginning of the period; shorter than that and you run the risk of not having enough time to adequately review all topics.

The total amount of time allotted for the ISP is influenced by your school and what they will allow. It becomes complicated when students have to choose between their ISP and vacation. If you are only able to study for less than four weeks, make sure that you take this into account and increase your pre-ISP study significantly to account for the discrepancy. Remember, the USMLE Step 1 Exam is a score that you will have to live with for the rest of your immediate medical-training life. You have the rest of your life to take vacation and can make up for it in the fourth year (which is by some accounts, a year-long vacation). You will also have to balance the risk of burning out. We cover the infamous "burn-out" in our Motivation Strategy section. We describe signs of its impending arrival, how to prevent it, and what to do when it happens.

2.    Topics to be covered and study format

The following figures represent the major topic areas covered on the Boards. They are categorized into two formats: Systems-Based and Subject-Based. The Systems-Based list is separated into General Principles and Individual Systems. The General Principles are topic groups that are difficult to separate into systems given their fundamental nature within medicine (e.g. biochemistry, microbiology). In each of the Individual Systems, you will need to cover anatomy, pathology, pathophysiology, and physiology. Some students choose to include pharmacology in the Systems-Based schedule. You will need to determine which of the two study plans you would like to follow. The decision as to which schedule to follow tends to rest on which format resembles your school curriculum. Students tend to stick with the format with which they are most comfortable. Either way, you must divide up your total number of days among various topics based on the following considerations.

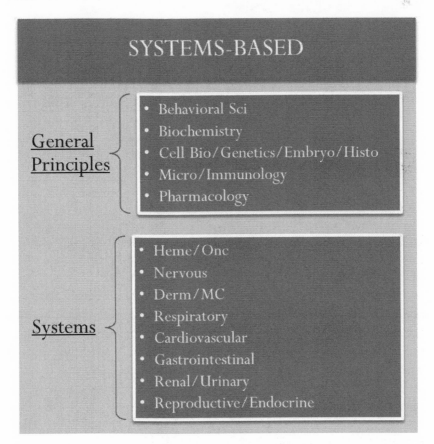

### SYSTEMS-BASED

**General Principles**
- Behavioral Sci
- Biochemistry
- Cell Bio/Genetics/Embryo/Histo
- Micro/Immunology
- Pharmacology

**Systems**
- Heme/Onc
- Nervous
- Derm/MC
- Respiratory
- Cardiovascular
- Gastrointestinal
- Renal/Urinary
- Reproductive/Endocrine

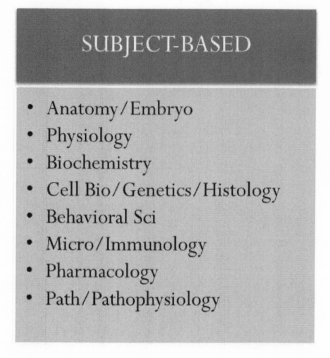

*Figure 6 Systems and Subject Based Organization of topics*

# Weighting each topic

To determine how many days you should spend on each topic, you must weight topics based on a few factors. We will cover the following areas that will dictate how much time you spend on an individual topic relative to others.

3.    Relevance to the USMLE Step 1 Exam

The first factor will be relevance to the Boards. This refers to what percentage of the test each topic will represent. The following is a rough estimation of what you can expect from the test. As you'll see, anatomy is covered the least, and pathology and pathophysiology are covered the most. You must keep this in mind when dividing up the number of days. You can spend whole or half days on individual topic areas (e.g. 3 or 3.5 days). This relevance data is not released by the NBME, and thus the following values are approximations based on our experience.

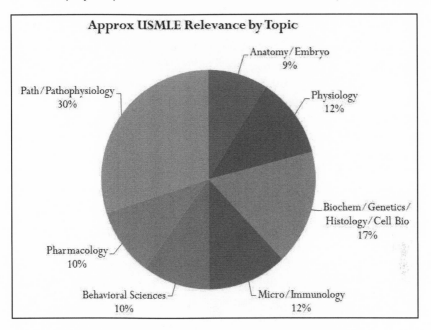

*Figure 7 Estimation of USMLE relevance by Subject*

4.  The amount of work to be done: Number of questions and additional reading

The number of questions in the question bank on each topic will largely determine the amount of work to be done. During your Intensive Study Period, you should expect to complete on average two, 46-question blocks per day. If a topic has 200 questions, you should allocate at least two days to finish that topic. If you plan to use a supplemental source in addition to your questions and central text, you should account for this by adding more days.

When reading a supplemental source, you should practice skimming and reviewing and not reading line by line. A large text like a BRS that is 300 pages should not take you longer than one to two days to finish. When using a supplemental source, don't read in isolation; alternate the reading with questions in two- to three-hour blocks. For example, if you reserve four days for the biochemistry, et al. section, you can complete the 280 questions and the reading over a course of four days. If you do decide to use supplemental sources for certain subjects, choose these texts prior to beginning your Intensive Study Period.

5.    Strength level

The final factor to consider when weighting topics in your schedule is your strength level in each topic. This comes into play when attacking your weak areas. If you know that you are particularly weak in an area, you will be more likely to use a supplemental source. In areas of weakness, you are also more likely to have more incorrect questions to rework in your question bank. For these reasons, you should allocate slightly more time to weaknesses.

You must balance apparent weakness in a topic group with the USMLE relevance factor previously addressed. If you are weak in a topic that has low relevance to the USMLE Step 1 Exam, you should not spend an exorbitant amount of time on it. A common example is that a majority of students consider themselves to be weak in anatomy and embryology. Students could probably spend the entire Intensive Study Period on these remote subjects and still not become experts. But given that these are the two topics that are tested least by the NBME, students should not spend the least amount of time in these areas.

6.    Chronological order of topics

An important factor to consider is the order in which you will place topics in your Intensive Study Period. You should aim to attack your weak areas first as these are the ones that need the most refreshing. For most students, these are the first-year courses (e.g. biochem, cell bio, genetics, behavioral sci, etc.). Your goal will be to eliminate any uncertainty or fear associated with these topics. By reviewing your weak areas first, you will have more opportunities for integration through the process of Free Association with subsequent topics.

7.    Practice tests

We cover the full NBME Practice Strategy at length later in the text but will briefly outline it here. You should take an NBME CBSSA (practice exam) prior to beginning your Intensive Study Period. You will want to allocate two separate days to take two NBME practice tests within your ISP. The first NBME practice test should take place at a convenient halfway point in your ISP. By this point, you should have attacked a majority of your weaknesses. The second NBME practice test should take place at a point between one week out and three days out from your exam. By this point, you should have covered all topics. This last NBME practice test will serve as a diagnostic benchmark that will give you an indication of where you stand and where you are still leaving points on the table in the form of persistent weaknesses. After this last NBME CBSSA, you will begin your Comprehensive Review Period prior to taking your USMLE Step 1 Exam.

Some students choose to take a fourth NBME practice exam. Students for whom this makes sense are those who have high levels of test anxiety and feel they need more simulation to gauge speed and the stress level of the exam. Some students may not receive a passing score on their third NBME practice test. For these students, we absolutely insist that they take a fourth NBME practice exam. You should not take the real USMLE Step 1 Exam until you have received a passing NBME practice score. We fully cover the strategy behind this in our NBME Practice Strategy section.

# The *Step 1 Method* Study Schedule Wizard

To facilitate this process, we have created the *Step 1 Method* Study Schedule Wizard. This software tool takes all of the above factors into consideration to produce a customized study schedule. There are four key inputs:

1. The total number of days available to study
2. Strength levels on individual topics (Strength is measured on a scale of 1-5.)
3. The desired order in which you'd like to complete the topics

Once you have entered these inputs, the Schedule Wizard uses an algorithm to create a customized schedule for you. At this point, if you'd like to modify the number of days for individual topics, you can do this. The schedule can be saved or printed for further reference. We hope to soon be able to upload the output of the schedule into popular scheduling programs such as iCalendar and Outlook. Our students love the tool because the act of creating an ISP schedule historically took days to weeks to plan. The creation of this tool required countless hours of iterations, programming, and proofing. We know that the work on this project is far from done. We continue to update the tool to add features students want most. We appreciate your feedback and ideas on how we can make the tool better. You can find a link to both systems- and subject-based study schedules on our online course.

## Executing the Intensive Study Period

Once you have planned your Intensive Study Period, you must carry it out. The Schedule Wizard will output a schedule of different blocks of topics on consecutive days, but it won't dictate how you spend your 8–10 hours per day studying. As mentioned, the majority of your time during the ISP should be spent working questions in a regimented fashion. Your basic goal for each topic group segment will be to get through all of the questions in the Qbank and to work through your incorrect questions twice. If you are not reading a supplemental source, your goal should be to finish two blocks of 46 questions a day. That's roughly 90–100 questions per day. If you are going through questions a second time, your speed will increase, and you might be able to do more than this.

When beginning a topic group, you should outline how many questions exist and make a realistic judgment of whether or not you'll finish. If you have to choose between getting through all the questions in the Qbank and working your incorrect questions twice, definitely choose to get through all the questions. Hopefully you will begin working through questions early and won't be faced with this difficult decision.

## How to Use Supplemental Sources

If you have a weak area for which you want to use a supplemental source, mix the reading with questions. Never read for more than two hours in isolation without working questions. For example, you may start reading the supplemental source for two hours, then stop and do a block of questions, then return to reading the supplemental source for one to two hours, then back to questions. Another way to use the supplemental source is as a reference text. Instead of planning to read through the entire text, you can have it on hand when you need clarification on a particular topic. You can read through that section and other surrounding sections to provide context for the general area. As expressed in the Principle of Free Association, when students have a weakness in a particular area, the surrounding areas are also weak. The NBME understands this and includes similar topics and themes as distractors when creating wrong answer choices. The NBME's goal is to probe how deep a student's knowledge base goes.

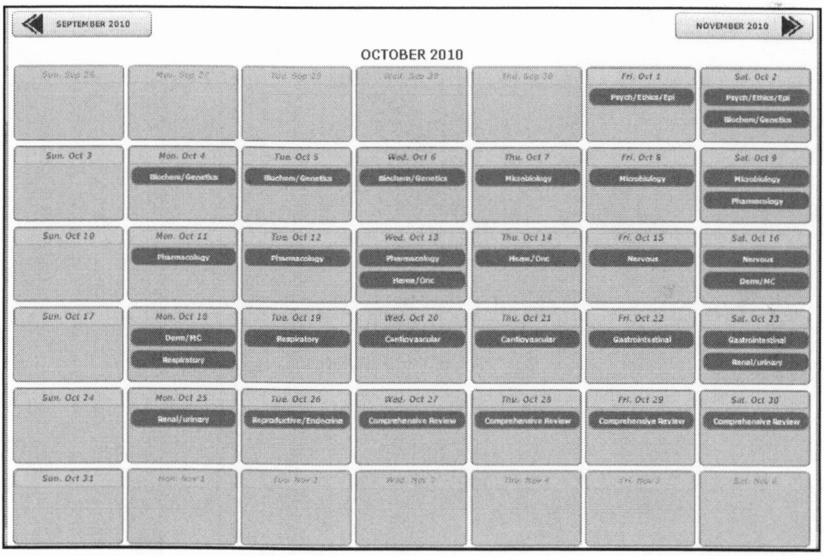

*Figure 8 Sample Study Schedule Output of Schedule Wizard*

# How Deep Do You Go?

Always work questions in the *Step 1 Method* format. Always think about the framework for each topic you are covering. Practice the Principle of Free Association as often as possible for as many topics as possible. Feeling the need to rush is a common problem students face during their ISP. They see a mountain of questions and attempt to get through them as quickly as possible. Fight this urge; don't rush the process. Your ISP is the time to digest and learn the material you didn't during the first two years of med school. You should struggle with your areas of weakness until you've understood the key elements and branch points of that topic. You want to do this in a timely process and make sure you are diving into the right weaknesses. As we addressed earlier, some weaknesses are low-yield topics and not covered extensively on the USMLE Step 1 Exam. Further, you want to make sure you are not spending longer than 20–30 minutes on individual topics, but if that's what it takes, so be it.

# Daily Study Schedule

Creating a daily schedule will be largely individual, but the fact is that no matter how you schedule it, you should be studying at least eight hours a day. We believe the optimal amount of time to study per day is 8–10 hours. Make sure to fit in time for frequent breaks, meals, and time to exercise or to get outside and get fresh air. At least once a week, you should go out with friends or do something completely unrelated to school.

Regardless of the format, create a daily schedule with goals that state the amount of work you will complete that day, and stick to it. If in the event that you do not finish the required amount, readjust the next day's schedule to compensate. When scheduling your exam, be generous to yourself, knowing that you will not always be able to finish all the work and may need extra time you didn't foresee when you initially laid out your plan.

I've included my daily schedule, complete with the activities that suited my lifestyle and interests. I am often a late sleeper, enjoy sleeping in, and enjoy going to the gym first thing in the morning. So I adjusted my schedule to reflect those personal, individual traits. That is probably the best thing about studying for the Boards: after having your schedule dictated to you for the past two years, you now get to make up your own schedule and learn at your own pace.

| Time | Activity | Time | Activity |
|---|---|---|---|
| 9:00a | Wake Up/Eat | 6:00–6:30p | Break |
| 10:00a | Gym | 6:30–9:00p | Study |
| 12:00p | Go to School | 9:00p–9:30p | Dinner |
| 12:30p–3:00p | Study | 9:30–11:30p | Study |
| 3:00p–4:00p | Lunch | 11:30p–12a | Go Home |
| 4:00p–6:00p | Study | 9 Hours of Studying | |

*Figure 9 Sample Daily Study Schedule*

# How Many Days a Week?

Some people preach studying only six days a week and taking a seventh day off for rest. While I was studying for the Boards, it did not make sense to me to have a whole day in which I did not study for the biggest test of my life. Thus, I studied all seven days, but once or twice a week in the evening I would go out with friends for a night out on the town. The great thing about this method is that you will cherish these nights out with friends or family, and they will sustain you throughout your study period. These brief respites break the monotony of your study space and the computer screen. Taking breaks will keep you fresh and leaving you wanting more from your Board study instead of dreading it, which can occur if you overdo it.

# When to Reset the Qbank

A common question we receive from students is regarding re-setting the QBank prior to their ISP (Intensive Study Period). Normally, you would wait until you complete 100% of your Qbank prior to resetting. Students will often complete 70-80% of their question bank prior to beginning their ISP, and they wonder if they should reset the question bank before starting their Intensive Review. Below, we cover the different issues students should consider when making this decision.

# The Easy Solution

The easy solution is to arrange your intensive study period such that the first topics you cover are the questions you have not covered yet. This is a good solution if the "uncovered" topics are weak areas for you. Remember, we want you to tackle your weak areas first before taking the 2nd NBME practice test.

You want to spend your intensive study period sticking to a schedule in attacking systems/subjects one by one. You do not want to spend the first part of your ISP doing random questions just to finish the QBank.

Thus, if you start your ISP with a subject that you have not yet covered, for example Biochem, then it will be a non-issue. You will do those questions in their entirety, and you will do your incorrect questions twice to make sure you cover the holes in your knowledge base. This will reduce your total % unfinished, and you will be one step closer to the magical "100% completion."

When you complete these unfinished topics, then you can reset the QBank and begin to go through the covered topics as if it were the first time. You will do these questions once through, then you will attempt to re-do your incorrect questions a second time.

# The Reality and the Complex Solution

If things were only this easy. I understand that many students have done questions in a "hodgepodge" fashion, with erratic gaps: they have 60 cardio questions left, 40 pulmonary left, 30 renal questions left, etc.

Now, if the unfinished questions happen to fall into topics that you will not cover until the 2nd half of your ISP, that means you will have to go through subjects you have already covered first. Doing questions in subjects you have already covered can be a frustrating experience if you do not reset the question bank. In order to see questions you have not done yet, you will select "unused" questions first, and then you will cover "incorrect" questions. There still may be a large portion of questions that you got correct at one point, but would like to review again. In order to see these questions again, you will have to create blocks of 46 questions in the "All questions" category and it may take a while to see all the questions because many will repeat along the way. Thus, you may see some questions 3 times, before you see other questions twice.

Thus, if you find yourself in this situation where you will be covering a significant number of topics you have already covered or partly covered, it can make sense to reset the question bank. Even if there is a small % of questions you haven't seen yet, I believe the benefits outweigh the risks if you know for certain you will be able to get through all of the questions in the question bank in the areas you have not covered yet.

Again, as you can see, the easy solution of covering areas with the unfinished % first, is a whole lot easier.

# The risk of resetting the QBank before finishing

Your risk is that for the % of questions you haven't seen yet, there will be some questions you will only complete once. How does this work? Consider that you have 15% of questions remaining, and for example, they are all in Cardio and Pulmonary. If you reset the QBank, you will still have not seen those questions. When you finally get to them, you will do them once through, then you will do your incorrect questions twice. The questions you got right the first time, you will not see again, because they will not be "incorrect."

This is not the case for questions that you had done prior to beginning your ISP. These questions you got correct, you will see twice. When you reset the QBank, you will see these correct questions again for the first time.

This is not a major problem if you make sure that in these "first time" sections (i.e. questions you will be seeing for the first time, and perhaps the only time) you are making sure that you are doing questions in the S1M format, reviewing the frameworks, and are getting questions right for the right reasons. Don't rush through these questions that you get right, because you may never see them again.

# Can you finish the QBank?

One main concern for students when they reset the QBank is that they will not have enough time to finish all the questions that they have remaining. To allay this concern, do some simple math. You should be able to do 2 blocks of questions (46 per block) per day on average. This is an average/minimum, you can do more than this; if you are re-doing a section you have done prior, then chances are you will be able to go faster than two blocks of questions per day. There may be days when you are using a supplemental source and will only get through 1+ block of questions on a particular day.

Thus, in a week, at best, you should be able to 644 questions or 14 blocks of questions. If you assume that the question bank is made up of 2200 questions or roughly 48 blocks of questions, then it should take you 3.5 weeks or 25 days to finish the entire question bank. This will still leave a few days for practice tests and comprehensive review if you have 28 - 30 days to study for the test. Hopefully you are not on this type of cramped 4 week schedule starting the QBank fresh, but it can be done, thus you should not be that worried about time.

Planning is the key to success in studying for the Boards. Check out our Motivation Module for more on the psychology of studying for the Boards.

# Confused Yet?

I understand that this can be a very complicated strategy but it is not meant to confuse you. As long as you understand the risks and benefits of an option, you can make an informed decision. The most important thing is to spend a majority of your time doing questions in the S1M format and learning the S1M Framework for each topic and question. You should be practicing the Principal of Free Association to learn similar categories of topics in concert to focus on similarities and differences between similar topics because this is where the Boards examiners will get their questions from. If you are able to do questions with these principals in mind, and are able to get through the question bank at least once, you will have a solid fund of knowledge going into the test.

# Chapter 11

## The *Step 1 Method* Practice Test Strategy

### The Role of Practice Tests

The question bank is too valuable an educational tool to be used as an evaluation tool. Many students look to the question bank percentages for the regular reassurance that they are doing well. If they don't receive this reassurance, they can become dejected and change their strategy unnecessarily. On the other hand, if students do receive high percentages, they can be lulled into false complacency. Students should use question banks as a tool to learn the information, and then use diagnostic evaluation tools to determine if they've learned the information. The evaluation tool of choice is the NBME Comprehensive Basic Science Self Assessments (CBSSA), also known as NBME practice tests.

### Diagnostic Practice Tests

The NBME CBSSAs are the only true diagnostic practice test on the market. These are the only tests that will give you an accurate representation of what your USMLE Step 1 score would have been had you taken the Step 1 Exam on that day. Studies have shown that the NBME practice test scores correlate at a high level with actual Step 1 scores.[10] We have recreated these results with our

---

[10]  Relationship Between Performance on the NBME Comprehensive Basic Sciences Self-Assessment and USMLE Step 1 for U.S. and Canadian Medical School Students. Academic Medicine. 85(10):S98S101, October 2010.

own student database scores. Many question banks will attempt to sell you practice tests along with their question banks. You should not buy these for diagnostic purposes. Many students mistakenly purchase these because they believe it to be a good value or the question bank providers create a purchase package discount.

If you have purchased question bank practice tests, you can use them as a way to simulate the test experience. You can practice your skipping strategy to ensure that you will complete 46 questions within an hour. But the score report that is produced by the question bank practice tests means nothing. Many students mistakenly believe that a great score on a Qbank practice test will ensure a great score on the actual USMLE Step 1 Exam. This is not the case. The NBME creates the CBSSAs by using the same formula of questions of different topics, subjects, systems, and styles used to make the USMLE Step 1. Further, the questions used are often retired or recycled USMLE Step 1 questions. The question bank providers do not have the USMLE Step 1 formula, and therefore cannot truly create a diagnostic exam.

The NBME CBSSAs are four blocks of 50 questions. You are given 1 hour and 5 minutes to complete each block. This time allotment creates the same scenario as on the USMLE Step 1 Exam: 1 minute and 18 seconds per question. The USMLE Step 1 Exam used to be 50 questions per block; thus I suppose that the number of questions in the CBSSAs is a reflection of that legacy. You can choose to take the CBSSAs timed or self-paced. If you take them self-paced, then you have up to four hours to finish each block.

# Cost

Each CBSSA costs $60 with expanded feedback and $50 without expanded feedback. Expanded feedback will show you which questions you missed on the exam. It will also tell you the average amount of time spent on each question by topic. We recommend that all students purchase the expanded feedback. All students will receive the standard score report with strength bars. This score report will tell you where you are strong and weak. But many times students will be told they are weak in a topic in which they had previously felt strong.

The expanded feedback will show the student exactly which questions they got wrong. This way, students will know why they are weak in particular areas. Further, the students can then remediate those weaknesses by searching through their resources to find the answer. Without the expanded feedback, students will find it difficult to tangibly improve their weaknesses.

Currently, there are six total NBME CBSSAs: five with expanded feedback (13, 12, 11, 7, 6) and one without (5). The CBSSAs are numbered in order of latest release: 13 is the most recent form and was released in March of 2012. The USMLE Step 1 Exam has changed significantly over the last 10 years. Thus the more recent CBSSAs reflect the most current version of the exam. As you'll see in our practice test strategy, we recommend taking the most recent CBSSAs. You can find out more information about the CBSSAs and see sample feedback forms at the NBME CBSSA website: http://www.nbme.org/students/sas/sas.html.

The NBME will also send out a free "tutorial" CD with four 50-question blocks—the equivalent of one practice test—for free in the introductory CD. These tend to be retired USMLE Step 1 questions and are in the same format as the actual exam. After completing each block, you will receive a percentage correct, but you will not be able to see the right or wrong answers. This reinforces the fact that the CBSSAs with expanded feedback are the only diagnostic tests that can help you target and improve your weaknesses.

## Simulate the Test

For diagnostic purposes, we believe you should time the practice tests to simulate the USMLE Step 1 Exam. When doing the practice tests, you should simulate a test environment and do each block of 50 questions in the allotted 1 hour and 5 minutes. Simulate the experience by taking the exam in a secluded location away from distractions and with secure Internet access. Plan to take a 5–10 minute break in between blocks just like you will in the real exam. Some students want to take an eight-hour practice test to truly mirror the eight-hour USMLE exam day. A four-hour exam is enough of a simulation to serve this purpose. Remember, practice exams can only teach you so much and are diagnostic tools more than anything else. It is better to reserve the second half of your day for learning and attacking weaknesses.

# The *Step 1 Method* Practice Test Strategy

The Practice Test Strategy is meant to provide regular evaluation of your study efforts. We believe that you should not evaluate yourself before you have learned the information. That said, most students would always prefer more time to study. The extreme of this sentiment is seen in students who are too afraid to take a practice test for fear of not being ready. They will go throughout their entire ISP without taking a practice test. They will take one shortly before their exam and realize that their score is far away from where they want it to be. With their exam only days away, they don't have time to readjust what they're doing or to focus on a particular weak area. The other extreme is students who take too many practice tests. These students will take a practice test every week. They are surprised when their score does not move as much as they would like. They are often shocked when their score actually goes down in this setting. In our Practice Test Strategy, we try to find a balance between regular monitoring and time for actual improvement.

# The First NBME Practice Test

You should aim to take your first NBME practice test at the end of your second year or before you begin your ISP. The theory is that you should have seen a majority of all topics on the exam. If you've been practicing the *Step 1 Method*, you should also have completed a significant portion of your question bank. This will give you familiarity with the USMLE style of questions. You will also have started studying in a way to learn information at a deep level of understanding. Although you won't have finished all of the Qbank or reached all topics in *Step 1 Method* style, you will have begun the process. Again, make sure to receive the expanded feedback option so that you can effectively target your weaknesses.

Many medical schools are starting to administer CBSSAs at the end of the second year. You can use this as your first practice test. It may be unclear which CBSSA they are using, so make sure to inquire which form they are administering. You want to make sure that you do not retake this form during your ISP. Many students have asked whether or not it is okay to retake a CBSSA that they've already taken in the past. This test will no longer be diagnostic if you take it again because you have already seen the questions. Even if you don't remember the answers, the questions themselves will have entered your consciousness, and since the original exam, you may have been subconsciously looking for the answers. Many students who have to retake the USMLE Step 1 Exam find themselves in this situation. They have taken all of the available CBSSAs and are now left with few accurate diagnostic markers.

After taking the first CBSSA, you should take a look at the score report of your strengths and weaknesses. See if it makes sense. Usually, most students will be weakest in the areas that they have not covered recently. Usually these are the first-year topics (biochemistry, cell bio/genetics, anatomy/embryology, behavioral sciences, physiology, etc.). Other weaknesses may include second-year, system-based topics that you never fully grasped. For example, neuro, immunology, and heme/lymph are topics that can be difficult to grasp if curriculums don't dedicate a significant amount of time to them. Finally, everyone has their own individual topics that have been persistently weak throughout the second year. Many students are weak in pharmacology, and this will be clearly shown in the score report. Next, take some time to look at the expanded feedback. See what questions you missed that correspond with weak areas. At this point, you do not need to look up the answers, but you should note on what topics you missed questions. When you reach that topic group during your ISP, you will need to learn about these topics.

One of the biggest mistakes we see students making is that they will make the same mistakes repeatedly. It is not because they are error prone, but rather they have not addressed their weaknesses. The NBME is notorious for choosing a handful of high-yield topics and asking multiple questions about these topics on an individual exam. If you get four questions on congenital adrenal hyperplasia on your exam, and endocrinology/reproductive has been a persistent weakness for you, you can kiss those points goodbye. The goal of these NBME practice tests is not only to identify weaknesses; you should also be correcting them in the process. This is the only way that you can reach your potential on this exam.

After surveying your strengths and weaknesses, you should use this information to plan your ISP. As we noted in scheduling your ISP, you should make sure that you devote more time to weaknesses. You may want to review a supplementary source during this time or you may end up having more incorrect Qbank questions to redo in these areas. Refer back to that section for more on incorporating strength level into scheduling your ISP.

You should aim to attack your weakest areas in the first half of your ISP. These are the areas that need the most attention and review. By covering them first, you'll have more time dedicated to them. Many times students try and save weaknesses until the end so that they can be "fresh." Most students tend to run out of time in their ISP as everything usually takes longer than you would like. You don't want to run out of time while covering your weaknesses. You'll also be able to review weak areas more frequently by connecting similar themes with subsequent topics through the Principle of Free Association.

# The Second NBME Practice Test

After covering a majority of your weak areas, you should be near the halfway point of your ISP. It is at this point you should take a second NBME practice test. Prior to taking the NBME, you should undertake what we call "pre-NBME review." By this point, you have covered your weaknesses and will be well versed in these. You will not have recently seen some of your stronger areas in some time; thus these will turn into relative weaknesses. To rectify this, prior to taking the second NBME, you should spend a few hours reviewing topics that you have not yet covered. Your goal is simply to refresh your mind that these topics exist, and a quick review through the *First Aid* is sufficient. Most of these topics should be second-year topics covered in the last half of the year. Thus, hopefully you've worked several questions in these areas and have annotated the *First Aid*. Simply reviewing text and annotations will jog your mind back to individual questions. This sort of review will be adequate for pre-NBME review. You will repeat this pre-NBME review routine prior to your third NBME practice test as well. Finally, your Comprehensive Review

Period prior to the real USMLE Step 1 Exam is a slight derivative to the pre-NBME review.

Again, many schools are now purchasing CBSSA vouchers for their students to purchase these exams. This is great, but make sure that the school is purchasing these vouchers to include the expanded feedback option. As we've shown, the expanded feedback option is a crucial part of the CBSSA.

After receiving the results of your second NBME practice test, go through the same process of reviewing the score report and incorrect questions. If the incorrect questions fell into topics you have already reviewed, immediately search for the answers and review the surrounding Free Association areas. If the incorrect questions are in topics not yet covered, you can take note of them and cover the areas when you get to them. It's important not to overlook these missed areas, so as not to make the same mistakes again. Remember, the NBME does not write new questions every year. The examiners take old questions and recycle them, covering the same topics in a slightly different form. Thus, the questions you are seeing on the NBMEs will often be revamped and could show up on your USMLE Step 1.

# Practice Test Analysis

Now that you have two score reports, you should begin what we call practice test analysis. You should compare the two score reports and pay attention to the differences. You should ensure that your weaknesses have improved since the last exam. If not, you should understand why not. The expanded feedback will show which questions you missed that fell into that category. Often, your previous strengths may have fallen. This is understandable because you have not covered them recently. When strengths fall and weaknesses get stronger, the total score may not change all that much. Do not become frustrated when this happens. This is actually a good sign. Your strengths will quickly become strengths again after you review them. It is much more difficult to improve persistent weaknesses. We've included an example of a real student's practice test analysis in the following section.

Many students become discouraged when their second NBME practice test score is not significantly higher than their first. They may tend to unnecessarily change their strategy. Do not get concerned unless it is required. As mentioned, the only time to get concerned is when persistent weaknesses have not improved. Even still, you will be able to see which questions you are missing. By and large, the reason why students do not improve persistent weaknesses is that their knowledge base does not go deep enough. These students tend not to be learning the entire frameworks for topics and they are not practicing Free Association. The NBME is very good at asking questions just out of reach of your level of understanding. It is a constant battle to make sure that your level of understanding goes deep enough to satisfy the expectation of the NBME.

## The Third NBME Practice Test

After reviewing the final topic groups, you should plan to take a third NBME practice test about five to seven days out from your exam. Again, you should perform the pre-NBME review prior to taking this exam. This time, you will likely be reviewing topics from the first half of the ISP. These prior weaknesses will now be remote and it is important to refresh them going into the CBSSA. The result of this last CBSSA will most closely reflect your performance on the USMLE Step 1 Exam. Again, make sure to obtain the expanded feedback so that you can learn from this exam.

Perform practice test analysis on this score report just as you did with the second NBME practice test. We recommend that you place all three practice tests below each other on a word document to compare the trends, which we illustrate a little farther on. This will allow you to evaluate persistent weaknesses.

## Reaching Your Potential

It is much easier to move a topic from the lower performance category to the borderline performance, than it is to move a topic from borderline performance to higher performance. In the same light, it is easier to move a topic from borderline performance to high performance than it is to move a topic from high performance to higher performance. The NBME purposely stratifies the difficulty of its questions in each topic group. You should not miss questions they expect most students to get correct. With adequate review of your question bank and central text, you should get these questions correct. The NBME is not trying to trick you with these questions. The higher performance questions, on the other hand, are less straightforward. You can be well versed in a topic and still miss these questions if you were not able to follow the NBME's thought process. You never want to leave easy points on the table. The third NBME score report is your final benchmark to signify where you are leaving easy points on the table.

# The Importance of Expanded Feedback

We cannot underestimate the importance of obtaining expanded feedback when purchasing the CBSSAs. It costs an extra $10 per test, but it is well worth the investment. Expanded feedback allows you to target your weaknesses more effectively. It will tell you exactly which questions you missed in a particular topic group. Without expanded feedback, the NBME score report will give you relative strength levels in all topics but will not tell you much more. If a student wants to remediate this weakness, they will be forced to initiate a broad, non-directed strategy. For example, if they are told they are weak in biochemistry, they will have to attempt to review all of biochemistry. This review will likely take several days, and they may not need to review all of biochemistry. Had they obtained the expanded feedback, they would have realized that they missed multiple questions on lysosomal storage diseases and protein structure and function. They then could attack these areas in a high-yield fashion and ensure improvement. Unfortunately, the broad-review strategy almost prevents you from diving deep into the actual areas where you need to review the most.

Expanded feedback also prevents students from repeating the same mistakes. Often, students will take multiple CBSSAs during their ISP and find that many topic groups are persistently weak despite review. One reason why this may be is that the NBME prefers to recycle old questions and topics to create new questions.

This is opposed to the more time-consuming and costly task of writing 10,000 new questions every year. Thus, if students do not learn why they are missing questions in a particular topic area, they run the very real risk of missing a similar question on a future exam.

# Sample Practice Test Analysis (Interpreting the Results of NBME Score Reports)

We've included a real student's practice test analysis. You can see how certain topics will undulate throughout the ISP. You have to understand that this is the nature of identifying and targeting weaknesses in an iterative fashion. It is to be expected. Never get discouraged in the middle of the process. You should anticipate that you may forget certain topics. You have to learn to review periodically and not to get frustrated when you forget something. Remember, getting questions wrong before the test is good. You want to find all of your weaknesses before you walk into the test. The more you review a topic, the higher likelihood you have of mastering that topic. You should not try to just memorize topics; you should contemplate and anticipate how else the NBME can ask a question about a particular topic. What will the next question on this topic look like? If you find yourself consistently forgetting a particular topic, stop for a moment and contemplate the topic. Why do you keep forgetting it? Do you get it confused with a similar process? Is the detail too minute? Do you not understand the mechanism and pathogenesis? Thinking about the answers to these questions will invariably get you to a deeper level of understanding about the topic.

The CBSSA score report displays your performance in each subject/system on the test into three different performance categories (Lower, Borderline, and Higher Performance). The strength bars represent the spectrum of difficulty in questions that you are answering correctly or wrongly in each category of performance. Remember, the NBME purposely puts questions of differing levels of difficulty on the test to be able to separate scores. The right edge of the strength bar represents the most difficult question you got right; the left edge of the strength bar extends into the weakest region where you got questions incorrect.

If your strength bar extends into the higher performance category, that means you were getting questions in the higher performance category correct. If the left edge of the strength bar extends into the lower performance category, you got these easier questions wrong. Thus, if you answer questions right and wrong in all three categories, the strength bar can span all three categories.

# Practice Test Analysis

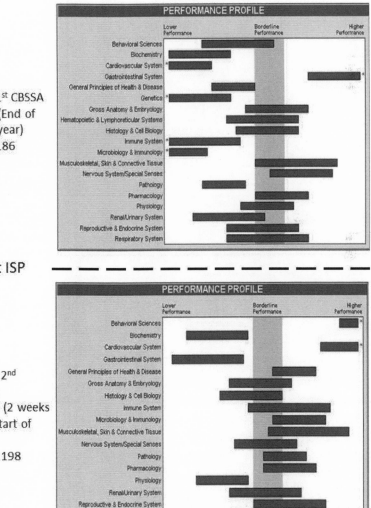

What: 1st CBSSA
When: (End of school year)
Score: 186

Start ISP

What: 2nd CBSSA
When: (2 weeks after start of ISP)
Score: 198

What: 3<sup>nd</sup> CBSSA
When: (4 weeks
after start of ISP)
Score: 214

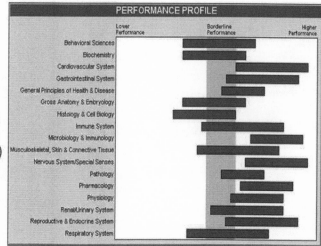

What: USMLE
Step 1 exam
When: (5 wks
after start of ISP)
Score: 223

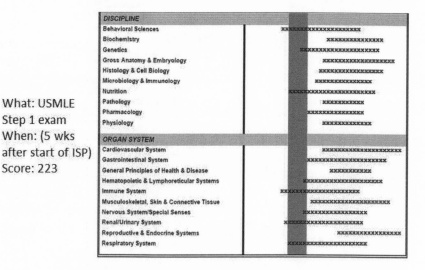

*Figure 10 Sample Practice Test Analysis and Score trajectory*

# The First CBSSA

In this student's practice test analysis, their first CBSSA is given to them by their school. They take the exam shortly before the end of the school year. They continue to study as they normally would.

The student is understandably concerned that her first score is not a passing score. Many students will be far away from a passing score on their first NBME. (Later in the chapter, we cover the concept of the first NBME exam and your score trajectory.) By this point, she has gone through a large portion of the Kaplan Qbank but has not worked questions in the *Step 1 Method* format. She plans to tackle the USMLE World question bank in her ISP and is motivated.

She sees that her main areas of weakness fall into the categories that fall predominantly in the lower performance category. In her case, they are biochem, genetics, cardio, micro/immuno, pathology, and general principles. She inputs this information into her Schedule Wizard and plans her ISP schedule. She chooses to attack first-year topics and major weaknesses in the first two weeks. She begins her ISP and takes a second CBSSA after two weeks.

# The Second CBSSA

After attacking her major weaknesses and first-year topics, the student has made substantial improvement. She realizes that biochemistry did not improve as much as she'd like. She'll continue to review and integrate biochemistry along the way. If there were incorrect questions she did not have time to repeat, she will try to get those in when possible. She also notices that a previous strength and weakness have switched spots. GI was a strength and cardio was a weakness on the first exam. She reviewed cardio in the first two weeks but hasn't reached GI yet. Not surprisingly, cardio is now a strength and GI is now a weakness. She is not worried about this, but understands that her total score would have been higher had she had a chance to do more pre-NBME review with GI.

She is happy that micro/immuno is no longer a major weakness. She realizes that physiology has decreased in strength. She is going through questions in a systems-based method, so this likely represents weakness in physio in certain systems. She is guessing that she probably missed a lot of GI, neuro, renal, and respiratory physio questions. This is a safe assumption to make. In her last two weeks of her ISP, she will attack these remaining systems, making sure that she grasps the physio concepts.

She is a little frustrated that her score has not yet broken 200. She knows that there are still major weaknesses she has not tackled. She also knows that she is getting through her USMLE World for the first time. She is a little disappointed that she is not able to get through all of her incorrect questions a second time. She understands that this is what happens when you choose to do two different Qbanks. She comes to terms with this fact and resolves to do the best that she can with the time she has remaining.

# The Third CBSSA

In the last two weeks of her ISP, the student attacks the remaining systems with special attention paid to her weakest areas. She is six days out from her real USMLE Step 1 Exam. The third CBSSA will tell her two main things: what general range her predicted score is in and where her weaknesses remain going into her Comprehensive Review Period (CRP). In her pre-NBME review, she reviews topics from the first half of her ISP because her brain is filled with topics from the last two weeks.

She is happy to see steady improvement in most areas and a score increase of 16 points from her last exam. She is happy to see that her major weaknesses from the last exam—biochemistry, GI, physio, and neuro—have improved. Going forward, she sees that her weaknesses are any topic that lies partially in the lower performance category. Namely these are biochemistry, histo/cell bio, gross anatomy, respiratory, and behavioral sciences. She also notices that behavioral science was previously a strength but has since been falling.

She keeps all these things in mind when planning her Comprehensive Review Period. As we describe later in the chapter, in the last five days before the exam, she will review all topics going into the test. She will spend a disproportionately large amount of this review time on her weaknesses. This will allow her to make sure she gets the easiest questions right and does not leave any points on the table. Remember, it is much easier to get topics out of the lower performance category than it is to raise higher performance topics.

# The USMLE Step 1 Exam

Prior to taking her USMLE Step 1 Exam, the student planned and executed her Comprehensive Review Period. She spent this time reviewing all topics, but she spent a majority of her time attacking her weakest areas. A look at her final USMLE Step 1 score shows this. She took the exam five days after her last NBME practice test. In those five days, she increased her score by another seven points by improving some of her weakest areas. These most improved topics were biochemistry, histo/cell biology, gross anatomy/embryology, and physiology. Many of her remaining topics retained the strength she expressed in her last NBME exam.

It is remarkable to think that in only five days one can increase their score by seven points. But for students who still have considerable weaknesses in the lower performance categories, these are the easiest points to obtain. These lower performance category questions are based on the "bread and butter" topics that can be readily learned if students go through their Qbanks and central texts in the *Step 1 Method*. Higher performance questions are less straightforward and often include experiment-based scenarios or esoteric topics.

In order to achieve this last-week push, the student also performed many intangible tasks well. We cover pre-test logistic prep later in the text, but this student got plenty of sleep the night before and planned out her break strategy ahead of time. These intangibles are essential to remaining refreshed and maintaining the best mindset throughout the test.

# NBME Exams and *Step 1 Method* Score Trajectory

The student in this example increased her score by 37 points between the first NBME exam and the USMLE Step 1 Exam. All students who study in an effective fashion should expect their scores to go up. What has been less clear is how much should a student's score go up between their first practice test taken at the end of the second year and the actual USMLE Step 1 Exam?

To answer this question, we've begun to research the phenomenon of score trajectory using our student data sets. Anecdotally, we've noticed that most of our students increase their score between 20 and 50 points by the time they take the actual exam. We've given significant thought as to what accounts for this range and have performed some statistical analysis to validate our ideas.

In general, we've found that your first NBME exam score has a lot to do with what your final USMLE Step 1 Exam score will be. In one data set, we found the correlation to be as high as $r = 0.75$. This means that you can predict 75 percent of the variability seen in USMLE Step 1 scores based on what a student scored on their first NBME practice exam. Practically speaking, the first NBME exam has little to do with the actual USMLE Step 1 Exam. The two tests are not linked, and high performance on one does not guarantee high performance on the other. That said, performance on the first NBME exam taken at the end of the second year is directly linked to other important factors that are related to performance on the USMLE Step 1 Exam.

We've found that performance on the first NBME practice exam and score trajectory is related to two major factors. If you are surprised or disappointed with your first NBME score, consider the following factors to make sense of the result and to plan effectively.

## 1.   Where you are in your prep process

Many students have not completed a significant portion of the Qbank or attacked many of their weak areas by the time they take their first CBSSA. Some students have already gone through a significant portion of the Qbank but have not incorporated the *Step 1 Method* style of working questions. This requires high-yield annotation and incorporating the *Step 1 Method* Framework into learning about topics. If you still have a considerable amount of work to do in this regard, you can expect to see an initial score at the lower end on the spectrum. At the same time, you should expect to see a much larger point increase if you complete these tasks.

Because of this, it is common for many students to not have passing scores on the first NBME practice test. Another reason for initial low performance is that many students do not do any pre-

NBME review. Many schools will offer the first NBME practice test on a Saturday, and students will just show up and take the test cold. Because many students have not seen first-year topics in over a year, it is not surprising that they score low in this scenario.

If you have already gone through a large majority of your Qbank in the *Step 1 Method* format, your score might already be in the 220s, above the national average. Thus, your room for improvement will be capped and the numerical amount of score improvement will be lower. More analysis on this will follow.

## 2.    The score on the initial NBME practice exam

It is much easier to go from a 180 to a 220, than it is to go from a 220 to a 260. Although both scenarios represent a 40-point increase, the trajectory of the increase is very different. In the lower range of scores, knowledge acquisition skills are tested. For example, can you learn the framework for COPD and answer a two- to three-step question about it? If you are able to do that for a majority of topics, you will score above average, likely in the 230s. In the higher range of scores, knowledge abstraction skills are tested. As we have discussed, these are more experiment-based questions and hypothetical scenarios that require you to abstract the knowledge you have learned. To score in the 240s or higher, you will have to master the knowledge acquisition skills and then have a significant grasp of the concepts to abstract the knowledge you have learned.

Thus, if your initial score is greater than 220, you may only see a 20–30-point increase in your score, despite studying very hard. These knowledge abstraction skills are much more difficult to acquire. Visit our website and online course for our latest efforts in developing modules to try and help students acquire these skills.

Some students understandably feel pressure when thinking about high scores. Your goal should be to simply reach your potential on this exam, whatever that is. The path to higher scores is the same. If you go through the *Step 1 Method* as we have described: attacking your weaknesses, becoming comfortable with the format, learning information at a fundamental level, and maintaining an optimal mindset going into the test, you will reach your potential. With proper, steady preparation, there is nothing that can stop you on this exam.

The real goal here is to acquire as much knowledge as you can to prepare for the wards to take care of patients. Your future patients will appreciate all the hard work you invested to learn about their many disease processes. Try and focus on this clinical element of your hard work and less about the test itself. The great thing about the *Step 1 Method* Framework is that this is how clinicians think and apply their knowledge. Fortunately, it is also how the NBME produces test questions.

# Chapter 12

## The Comprehensive Review Period

### The Comprehensive Review Period

The Comprehensive Review Period comprises the last days of your Intensive Study Period. Here, you will make one final pass through all topics in a strategic fashion. At this point, you either feel as if you are running out of time, or you'd like for the test to be over already. Either way, this time should be spent tying together ideas, attacking weaknesses, and getting your "game-face" on.

### What Is the Purpose of the Comprehensive Review Period?

You will have covered hundreds of individual topics by the time you walk into your test, many of which were covered at the beginning of the ISP or at the end of the second year. You want to walk into the test with all the information refreshed, accessible, and at your fingertips. The best way to do this is to attempt to review as much as you possibly can before the test. You also want to make sure that you spend a disproportionate amount of time on your weaknesses as diagnosed in your last NBME practice test.

You should spend at least 10–15 percent of the total time at the end of your Intensive Study Period doing comprehensive review. For a month-long study period, this should be between three and five days. The Comprehensive Review Period should be spent doing three things: practice tests, skimming your *First Aid*, and working questions in randomly grouped topic groups.

# Practice Tests

Depending on how much time you have, you want to do an NBME practice test at least one week prior to the USMLE Step 1 Exam and before beginning your CRP. For most students, this will be the third NBME practice test they have taken, and it will simulate the actual test in both score range and question format. Remember four hours of simulation is enough to simulate the entire test. Do not use eight-hour Qbank practice tests; they will just waste your entire day and they won't be diagnostic. The key at this point is to get you into the mindset of hunkering down knocking out 46 questions in 60 minutes, and doing it repeatedly. You should be taking the NBME practice exams in a timed and Step 1-simulated environment. You should be practicing your question strategy principles. As mentioned, the score report and the expanded feedback from this last NBME practice exam will serve as the basis for planning your CRP.

There are two groups of students who may benefit from taking two NBME practice tests during their CRP. The first group is students who struggle with test anxiety and timing. If you are still struggling with these issues during your third NBME practice test, it may be a good idea to take another NBME practice test two days before your actual exam. The second group of students who may benefit are those who are in the gray area of passing the exam. Remember, you should not take the exam if you have not yet passed an NBME practice exam. You should postpone your USMLE Step 1 Exam until you have successfully passed an NBME practice test.

Our studies have shown that the margin of error between the last NBME practice exam and the real USMLE Step 1 Exam is ±10 points. One internal data set showed that when students do worse on the USMLE Step 1 Exam when compared with their last NBME practice test, they score seven points lower on average. Thus, we recommend that you have a passing margin that accounts for the potential of scoring lower on the real exam. If the passing score is 188, we'd recommend having a score in the mid-190s before sitting for the exam.

Again, we've had students who find themselves in the gray area at the beginning of the CRP and attack their weaknesses. As a result, their USMLE Step 1 score may stay the same or go up. We've also had students whose score goes down slightly on the actual USMLE Step 1 Exam. These students are very grateful for the margin of error that they had. Performance on the USMLE Step 1 Exam involves several intangible factors that occur on test day that are sometimes out of a student's control. Sometimes students have a bad day on the exam and make mistakes they normally wouldn't. You can do your best to prevent this from happening by simulating the experience several times and being well prepared. You also want to make sure that there is some score wiggle room in case you need it.

## Skimming the *First Aid*

You should plan to skim through the entire *First Aid* (or central text of your choice) during your Comprehensive Review Period to take in all of the information you have covered. It is truly an amazing experience to look back at your central text and see your scribblings over the past several months. Instantly, your memory will be jogged back to questions that were the source of the information. Don't reread everything! You've already read the information; your goal is to refresh and remind yourself that these topics exist and to go through the frameworks in your mind. When faced with questions, your mind will access the relevant information. The refreshing piece is key; your goal is to remind your mind that those topics are available for its use. The best way to do this is in an organized fashion along with questions as we'll review next.

## Working Questions in the CRP

Many students worry about wanting to work questions in random topic groups early on. We discourage working random questions when you're learning information because it leaves you scatterbrained going from one new topic to another unrelated, new topic. You don't get the necessary level of repetition and coverage of difficult-to-grasp concepts in a short amount of time. The NBME practice tests provide good practice with working questions in a random format. In the CRP, we utilize a modified format of working random questions that will allow you to cover several topics at once for time's sake. Further, this will prompt you to be able to move from topic to topic prior to the exam.

Planning your CRP is similar to planning your ISP. You will want to determine how many days you have to review and divide the days into the different topic groups you have to go over. You should spend more time on your weaknesses than other topics. In the CRP, which is usually only around five days, a half day on an individual topic is a lot. We recommend isolating your weakest areas into individual blocks of three-hour to one-third days. You should read through your central text and work questions in your question bank. If you know there are specific topics that you need to review, hit those first. Your goal will be to get through the central text, mixed with 30–40 questions. You will want to go through as many questions as you can in the *Step 1 Method* format to reinforce the frameworks and see which individual topics are unclear to you. Utilizing the expanded feedback from the CBSSAs will be crucial in identifying which specific areas need review.

For the remaining topics, you should place them in groups of three with similar strength levels (e.g. cardio/pulm/renal or micro, biochem, pharm). Then, spend one to two hours skimming the *First Aid* in these topics and then do a random block of 20–30 questions in these areas. You can also complete questions and read the central text simultaneously in the *Step 1 Method* format. You should repeat this format, until you have finished the central text. This should take you at least two full days, if not more. As soon as you receive your last NBME exam results, plan out which topics you will place together and when you will do them.

# Chapter 13

## Days before the Exam and Exam Day

Days out from your test, the nervous energy will build up. You want to plan ahead for this time and use that energy for good and imagine the best possible outcome. If you've followed the *Step 1 Method*, you are prepared. You should go in confident and be able to swipe blows with the test, and you will walk out knowing that you gave it your best. Remember, all you can do is your best. You have prepared in the most efficient way possible. You have simulated the test experience for the past several months, and now you are finally ready to take on the real thing.

In this chapter, we cover the last-minute things you should think about going into your test—the night before and the logistics of test day (skipping strategy, break time, the test site, snacks/meals, and the optimal mind set).

### Get Plenty of Rest

The night before the test, get plenty of rest. Despite your concerns, the information on the test will represent the material you studied in the last six months, not the night before the exam. Pack in your books in the afternoon and try to relax in the evening. You will want to have a clear mind when you go into the test. You will have to recall things that you covered at the beginning of the Intensive Study Period. If your mind is cluttered with random topics studied the night before at 10:00 p.m., it will be very difficult to integrate across all of the topics. You want to wake up refreshed, as if you were going to go to your daily study site to work another seven blocks of questions. That's all the USMLE Step 1 Exam is—seven blocks of questions. You've been working blocks of questions for months now; you're pretty good at it. And look at the bright side: you won't have to go through the leg work of looking up the answer and annotating your central text during the real test. One of our goals in the *Step 1 Method* is to increase familiarity and comfort with the exam. If you feel comfortable on exam day, you will be able to think clearly and access all of the information you have learned.

# Practice the Skipping Strategy

You should anticipate that up to 30 percent of the questions on the USMLE Step 1 Exam will be very difficult, abstract questions. Anticipate that you will have to use your question-strategy tips to tackle these questions. The remaining questions will be relatively straightforward, and you will have a good chance of knowing the information and answering them correctly. To give yourself the best shot at a high score on this test, you will need to be able to answer all of these straightforward questions right. In order to do this, you will need to be able to see every one of these questions and have enough time to answer them. To do this, you will need to practice your skipping strategy for the most difficult questions that most students get wrong. You should anticipate practicing the skipping strategy on at least five questions per block. If you're not familiar with the skipping strategy, review that section of the text, or watch our online module on Question Strategy.

# Don't Get Psyched Out

If you go into the test anticipating difficult questions, and particularly difficult blocks of questions, you will not be thrown off when you see them. Many students are discouraged by these difficult patches. Their mindset then gets thrown off for questions that follow that they do have the ability to figure out because if you approach an easier question with a discouraged attitude, you will not be as confident or effective. If you think a question is hard, you are not alone; most people will think the question is hard. Always stay positive and motivated, and simply focus on doing your best.

# Plan Your Breaks

You should have a good idea of how you are going to use your break time. The NBME officially gives you 45 minutes of total break time throughout the test. You are allowed to take breaks in between question blocks. There is no limit to how long or how short your breaks can be as long as the total break time remains under the 45-minute allocation. If you leave the testing room, you must check back in. This process can take anywhere from 30 seconds to two minutes depending on the number of people checking back into the room at one time. Next, we provide some suggestions on how to schedule and maximize your break time.

When you walk into the test, the first 15 minutes is reserved for doing a tutorial on the USMLE Step 1 testing software. You can gain an extra 15 minutes by doing the online tutorial ahead of time and skipping through it during the test. If you just click through it, the extra minutes are added to your break time. The link for the online tutorial is here (http://www.usmle.org/practice-materials/). If you finish blocks of questions quicker than the allotted 60 minutes, this extra time is also added to your total break time. You need to balance this benefit of extra break time with the risk of going too fast. Remember, you will only have 60 minutes for each block of questions and this never changes. If you practice the skipping strategy, you should be doing okay with time. If you expect a total of 60 minutes of break time (15 extra minutes from the tutorial), you have plenty of time to use the restroom and relax in between blocks.

Everyone is different with respect to taking breaks. Some students like to do two blocks of questions at once, while other students like to take short breaks in between each block. Personally, I am a restless person and prefer to take a short, five-minute breather in between blocks, even if I feel I don't need it. Further, you should plan to take a lunch of 20–30 minutes at some point midway through your test. You can also expect that your stamina will wear out as the day goes on, so you may want to position longer breaks toward the end of the day. You should have a tentative break plan in your mind when you walk into the test. See below for a sample plan:

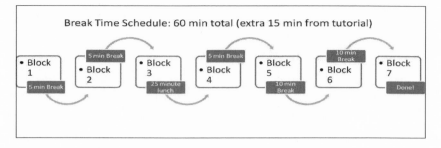

*Figure 11 Sample Break Schedule for Exam Day*

### Plan Your Snacks/Lunch

You should also plan to bring light, energy-producing snacks and beverages along with you to the test center. You should pack some sort of light lunch that can remain unrefrigerated (or bring a cooler pack). You don't want to bring heavy snacks or a heavy lunch that will weigh you down and put you into a food coma. Think fruits, nuts, power bars, crackers, etc. Think small sandwich for your lunch, not a large burrito.

Don't drink too many fluids or else you may have to use the restroom excessively. If you leave the test room mid-block, the clock will not stop. Thus it is better to use the restroom during breaks in between blocks. It's probably not a good idea to plan on buying your lunch because you cannot count on how long it will take for you to find something in the neighborhood. If you are a coffee drinker, definitely feel free to bring coffee. If you are not a coffee drinker, now is not the time to start. The general rule is to continue to do whatever has been working for you.

## Plan Your Travel/Documents

Do yourself a favor and figure out where your test center is before the morning of the test. Anticipate how long it will take you to get there through traffic or public transportation. Read the instructions on the test center printout to make sure you bring all the necessary forms of identification and testing documents. You don't want to deal with the stress of being late or forgetting something the morning of the test. Normally, they request that you get to the testing site 30 minutes before your set start time, but check your printout to confirm this.

# Just Do It

Your time is now! You've done everything we've asked of you. You are now adequately prepared to tackle this exam. Do your best, and leave it all on the "field." Your job is not finished until the seventh block is over. Do not give up before then, no matter how difficult you think the test is. If you see it through, you will be so glad that you did. Be proud of yourself and the effort you have put forth. You've done an amazing job up to this point; now it's time to finish the job. See you on the other side! Don't forget to write us and let us know how it went. Your feedback helps us make the program better every year.

# Section 5

# Motivational Strategy

# Chapter 14

## Managing Effectiveness and Stress

A majority of this text is spent on mastering the content and strategy behind the USMLE Exam. Still, reaching one's potential has a lot to do with reaching maximal personal effectiveness. By the time students reach the end of their second year of medical school, they have their preferred methods of studying. We caution students not to get too comfortable in their methods. The USMLE Step 1 Exam is very different from your medical school exams or the MCAT. It is a specialized test with a broad subject base and you are given only three to six weeks of dedicated time to study for it. In response to these extreme circumstances, you must ensure that all of your time yields high dividends. We have seen countless students who work very hard studying for the USMLE Step 1 Exam but do not get results that equate with the effort invested.

We've made the argument for why working questions in the *Step 1 Method* format is preferred. Working questions is an active, interesting way to learn the necessary material. Starting to work questions during your second year will require you to prioritize your activities and budget your time. You will have to invest a little more time on a weekly basis during the second half of your second year. In order to reach your potential, this is a commitment you will have to make.

We will examine subtle issues that can affect your effectiveness. These include learning to focus, group work, location, and managing distractions.

# Learning to Focus

Many students find it difficult to focus while studying for the Boards. This is a tumultuous time and there are many competing ideas and thoughts flying through your head. We will try to provide some recommendations on how to focus while studying and when taking the actual test.

# Focusing while Studying

Understand that the USMLE Step 1 Exam is a stressful time for everyone. If you find yourself getting anxious, understand this is normal but also understand that anxiety is a counter-productive emotion. Anxiety stems from worry that you are not going to be ready for the exam. Ironically, the time spent worrying will waste more time and make you more unprepared. Thus, when you find yourself worrying and anxious, realize it is because you want to do your best. The way to do your best is to focus and study on a daily basis. A great score is going to come from the hard work invested working question after question and learning topic after topic. Once you get into the mix of working questions, you will begin to forget your anxiety as you face the challenge of answering questions and learning.

If faced with continual thoughts of anxiety, focus on your score goal. Ask yourself what it will take to achieve your score goal. What sort of work, dedication, and persistence will be required for you to reach your potential? This sort of positive thinking and focus will also help you sidestep the persistent anxiety that many students face.

# Avoiding Distractions

By the time the Intensive Study Period arrives, most students have a set studying routine. Unfortunately, this routine may include constant distractions. The most common distractions are the cell phone, Internet, email, social networking sites, music with lyrics, and television. Although subtle, these distractions will reduce your effectiveness and steal 5–10 minutes frequently throughout your day. These distractions can add up to hours of decreased productivity in an individual day.

Distractions can be healthy rejuvenators if compartmentalized into scheduled break times. You should plan to study in two-hour blocks with 15 minutes of scheduled break time in between each studying block. You should shut out all distractors during this time (no email, no Internet other than used for studying, no social networking, no cell phone). Then during your break time, you should return calls, answer quick emails, and Google your favorite leisurely topics. A choice of music while studying can be difficult. Music with lyrics can be problematic as you may be prompted to sing along in your head. This may reduce your ability to dive deep into your weaknesses. Often when struggling with difficult topics, your mind will want to think out loud. You may find yourself talking through concepts and connections. This narrative may have to compete with the lyrics in your music. Listening to classical or light instrumental music in the background won't compete for mental processing space, and thus should be okay.

At the beginning of every day, you should schedule the amount of work you plan to finish. At the end of the day, you should survey whether or not you were successful. If you were not successful, you should honestly ask yourself two questions. First, were you too ambitious with your goals? If you are rushing through topics and not completely understanding topics in the hopes of finishing your desired amount of work, perhaps you should slow down and make more realistic goals. Second, if your goals are not too ambitious, are you not being as effective as you could be? If the answer to this question is yes, then you may be falling victim to many of the distractions we just discussed.

## Focusing during the Exam

Test day will be the most stressful day during your ISP. Some students perform better in stressful situations as their sympathetic nervous system heightens their senses and perception. Many students, however, may perform slightly worse in stressful situations as they feel pressured to make answer choices and cannot think clearly. One simple way to reduce stress while you are taking your exam is to focus on a few key thoughts. Hopefully you've implemented the *Step 1 Method*. If you have, understand that you have prepared well, attacked your weaknesses, and you are ready to perform. Your goal is to find all the questions to which you know the answers. You know how the NBME examiners think and how they like to ask questions. Implement your question strategy to work through questions as you have throughout your ISP. The actual exam is just another day of working questions. The only difference is that you have to work the questions at the Prometric test center.

Once you dive into questions, you will find that your mind is consumed with finding answers to questions and you will quickly forget how anxious you are. Anxiety can become rampant prior to beginning the test and in between question blocks. One way to combat this anxiety is through a practice known as deep-breathing meditation. The concept of meditation is meant to reduce the number of anxious thoughts flying through your mind. If you can "quiet" the mind, you allow more mental room for Free Association and content-based strategy that will get you to the right answer. The concept of meditation may be foreign and "weird" to some. I personally felt this way about meditation when I was first exposed to it as a second-year medical student. Once I let go of my preconceived notions about the practice, I began to implement it on a frequent basis.

To begin the practice of deep-breathing meditation, close your eyes while sitting comfortably. Take several deep breaths and focus on the air entering and exiting your lungs. Think about your diaphragm contracting and relaxing while controlling air flow into your lungs. Adjust the amount of air you inhale so that the process becomes natural and comfortable. Your diaphragm is innervated involuntarily, so it will contract on its own; perceive this process. As you will find, you can surprisingly think a lot about the simple process of breathing. What you will quickly find after 30 seconds or so is that you will lose focus on your breathing and your mind will wander back to your concerns, worries, and thoughts you had before you started the deep-breathing meditation. This is to be expected. The act of meditation is to be aware of your mind wandering and to refocus your thoughts back onto your breathing. The act of meditation is to repeat this process of refocusing when you find your mind wandering. What you will quickly see is that your mind incessantly wants to worry about particular things. After about three to four cycles of refocusing your thoughts back to your breathing, you will find yourself more relaxed and composed.

Remember, you are in control of your thoughts. If you don't want to be worried or anxious, you don't have to be. You can choose to be confident and accept your situation as it stands. You can choose to do your best and accept the consequences. If you have prepared well, I am confident you will be happy with the outcome. Your goal should be simply to do your best and reach your potential. The only way to accomplish this feat is to focus on answering questions to which you know the answers, and trying to figure out the answers to the questions you don't.

## To "Group" or Not to "Group"

Many people incorporate group study into their school-year study habits. There are both pros and cons to the idea of group study. Overall, we believe that group interaction is good if it is limited to a few specific types of interactions. You should never study solely in a group as this will lead to distractions and will cause you to sometimes review areas that are not targeted to your specific needs. Rather, we recommend periodic discussion sections where you go over topics that you have seen a lot or would like clarification on. These group interactions are best done over meal times such as lunches or dinners. Some students like to schedule weekly Board study meetings. At these meetings, students will work Qbank questions together or check in on other group members' study plans. Many students reach a deeper level of understanding when they explain concepts to others. In this way, you will not be sacrificing precious study time, and you will be passively learning on your break time. The implementation of these subtle techniques will take your understanding to the next level.

In addition, having frequent interaction with peers will prevent you from getting burned out and experiencing that "Boards Depression" that so many medical students experience. They say that misery loves company, and ironically it's true. Being around people who are going through the same thing you are will make it more bearable. Being able to share frustrations and gripes will allow you to cope and make it through an extended period of time to study.

# Location, Location, Location...

Borrowed from real estate lore, this saying holds true in the Board study realm too. You should have a primary study space for continuity that is quiet and will allow you to concentrate on your work with minimal distractions. Where do you normally study? Will you have to find a new location for intensive study? Think about this before you begin your Intensive Study Period. Interestingly enough, I spent most of my second year studying in the library and assumed that I would study there as well for my Intensive Study Period. I found, though, that most students in my class had the same idea, and the library became overcrowded and filled with multiple distractions. I then had to try out different locations until I found a location that allowed for optimal efficiency.

Evaluate your performance in your study environment. If you are not getting enough work done, change your location until you find a location that maximizes the amount you get done. I had to change my study location a few times until I found the optimal place, which was in my second-year classroom, ironically.

If you are going to be working question bank questions, you will need Internet access. Thus, make sure your ideal study spot has a reliable Internet connection. It is okay to change it up to a coffee shop or restaurant every once and a while for more leisurely studying and a pseudo-break. I would not practice this every day, but a few times a week for a change of pace allows you to break the monotony and be productive at the same time. Remember, you want to be enjoying your study time. If you are enjoying yourself, you will try to study as much as possible. Try and maintain this positive attitude throughout your study period to obtain optimal results. In the moment, it may seem like forever, but four to five weeks is a miniscule amount of time in the grand scheme of things.

# Chapter 15

## Managing Motivation

### Crossroads to Success

Invariably, all students will face major dilemmas during their Board study. It is best to face these dilemmas early on, as opposed to late in Board study because these thoughts of indecisiveness may compromise your effectiveness. The *Step 1 Method* requires you to face these mental challenges now so that when it comes time for execution during the intense month-long study period prior to your actual test, you will be prepared to execute.

### Decision Making

The difference between success and failure lies in the moments of the decision. You always have a choice. You can choose to fully apply yourself for this test and start studying as soon as possible by beginning prep work in your spare time. Or, you can choose to wait until the last few weeks before the exam to start studying, feel pressured, and have serious doubts going into the test. The choice is up to you. Remember you don't *have* to do anything—the *choice* is yours!

### Passion and Motivation

Why are you studying so hard for this test? Is there a particular specialty that you have dreamed of entering? Is there a particular program that you have dreamed of going to? Ask yourself these questions, and once you have the answers, focus on them. Visualize yourself achieving these goals, and see yourself in this future position. Become that person. How hard would this person work? How much medical knowledge would this person have? Start carrying yourself as if you were already this person; practice that work ethic today! Constantly remind yourself why you are studying for this test: put up pictures, posters, or other reminders that will constantly motivate you when your drive starts to wane.

## Persistence and Failure

Committing yourself to any daunting task will always present obstacles. Invariably, there will exist a time at which point you feel as though things are not going well, when you think you "might as well give up because there is no hope." The surprising thing is that at that point, you are making the most progress because you are actually treading into unknown territory, doing what you previously could not! Ironically, when you feel as though it's time to give up, that's when you should actually put your head down and push harder. Blind persistence and never giving up will always lead you to your goals. Failure can become a mindset if you let it. Always remember that if you maintain focus and continually reevaluate your approach, you will always be successful.

## Open-Mindedness

Many theories of success state that your level of success will be determined by how many options you have. Basically, if you have more options at your disposal, you will eventually find the best one. But in order to find the best option, you must step out of your comfort zone and try new techniques. That is the reason that you are consulting this program—so use it! The techniques in this program have stood the test of time and have been used by thousands of students with great results. The techniques taught in this program may be new to you, but you must have faith and let the results speak for themselves.

# Define Your Outcome

As you begin such a large undertaking, you must prepare yourself for the best- and worst-case scenarios. The beauty is that you have the opportunity to choose your outcome now.

Oftentimes you will hear people studying with the goal of "passing"; they are thus motivating themselves to simply avoid failure. That is not a motivating goal. Goals were created so that you would have something to shoot, dream, and strive for. You need a goal that will motivate you to work hard. Mediocrity does not demand hard work; therefore, you should never shoot for mediocrity or just to pass. If you study and use the techniques in the *Step 1 Method*, there is no question that you will pass. The question is: How well will you do?

Many of us, although we are clearly high achievers, have not yet reached our potential. Our best days are ahead of us. Now is the time to seize the moment and shoot for your dreams. You can do it! You must first believe and realize that the opportunity is right in front of you. Then, you must seize the moment and capture and implement the tools to achieve it. You will never get this chance again. I believe that regret is probably the worst feeling in the world because you are helpless. Do not regret that you did not give it your all.

# Avoiding the Infamous "Burn-out"

Studying for greater than four weeks consecutively, day-in and day-out, will definitely take a toll on your psyche. Therefore, it is imperative that you frequently refresh your mind with activities that do not focus on study. A simple 15-minute break doing something like talking to a friend, watching a sports highlight clip, or answering an email, can keep you going strong for another two hours. Make sure that you spend some time each day maintaining your health through exercise, healthy eating, and getting plenty of sleep.

Studying for the USMLE Step 1 Exam is a marathon, not a sprint. How you do in the first week is just as important as how you do in the last week. Be careful not to go "all out" in the first two weeks because you may burn-out and lose interest and concentration in the last two weeks. If you do not spend time doing other things and having fun, you will associate all sorts of negative emotions with studying for the Boards. This is where so many students falter in their preparation. You will often hear students remark that they just "want to get the test over with." What in turn occurs is that they cut corners and just go through the motions until the test is over and done with. Inevitably, when they get their score, they are disappointed and wish that they had spent more time focusing on the details rather than glossing over them. By constantly refreshing yourself with periodic breaks and occasional extracurricular activities, you will always be energized and ready to tackle difficult concepts. Hang in there and stay positive.

You should view your Board study period as the best time in your life. For the first time ever, you will be in control. You will call the shots. You will determine what you study, how you study, and when you study. You will determine when you take breaks and when you finish those last few questions. This is an amazing amount of responsibility, but after two years of medical school, you are able to make good choices. You know your strengths, and you know your weaknesses. You know when you study effectively and when you don't. No one knows you better than you; therefore, who better to determine your preparation for the most important test of your medical career?

# "The End of the World"

Even though we stress the importance of the test, we know that the USMLE Step 1 Exam is not the end of the world. Many students who do not do well on Step 1 can improve their standing with excellent clinical grades, good Step 2 scores, and outstanding research. The goal of the *Step 1 Method* is not to have you perseverate on the Boards for the rest of your life, but to give you the tools to reach your potential on this exam the first time through. We know that there is great, untapped potential in every medical student or else you wouldn't be in medical school. We want to give you the tools to maximize your effectiveness so that you can get the most out of your time and be successful. The principles taught in the *Step 1 Method* can be used for the Step 2 Exam, and more importantly, can be applied to other goal-directed activities with great effect.

All of the mental preparation tools are taken from the latest effectiveness, time-management, and motivational literature. These tips have been proven to be effective not only in people's lives, but they have been proven on this exam. Thousands of medical students across have applied the principles in the *Step 1 Method* and many of them have gotten into their top choices for residency—many of which are in ultra-competitive specialties as neurosurgery, orthopedics, radiology, and ophthalmology.

# Chapter 16

## Special Situations

### Delaying Your Exam

Many students find themselves in a situation where they are debating delaying their exam. In general, we always encourage students to take as much time as they feel they need within a scheduled test period. This refers to the time period that your school allows you to take the exam. The contrasting case is when a student must extend their testing period by petitioning their school administration. Extending the testing period carries with it third- and fourth-year ramifications. These ramifications will affect your selection for third- and fourth-year electives and other opportunities that will affect your residency application process. These ramifications make the decision to extend your testing period a serious one.

If you've been implementing the *Step 1 Method*, you should know exactly where you stand throughout your Intensive Study Period based on your NBME practice tests. You should also know how to attack and improve your weaknesses to get the best score possible. Many times, external events such as family or health issues will preclude a student from investing adequate time into study. If you find yourself removed from study for a week or greater and your practice scores are not where they need to be, you should consider extending your test date. You should always involve a faculty advisor when making this decision. If you did not receive passing scores on your last NBME practice test, you should always extend your study time until you receive passing practice scores. Going into the exam, it is also preferred to have a comfortable score buffer of 5–10 points above passing.

The tougher decision occurs when a student has a passing NBME practice score, but feels that they've not met their potential. You must be honest with yourself and ask why you believe this is. Have you not implemented the tips of the *Step 1 Method* completely? Have you taken shortcuts throughout your Intensive Study Period? Was your originally scheduled Intensive Study Period shorter than the prescribed four to five weeks? Did you have to repeat a course and therefore get a late start on your study? These questions should all get to one fundamental question: what will change if you extend your study period? Contrary to popular belief, extending the testing period is not always better. We've analyzed the pros and cons, and you will see those next. If you do extend your testing period, make sure that you mitigate the risks associated with extending your testing period.

The Cons:

- You are farther away from material learned in the first two years and the beginning of your Intensive Study Period. More time to learn also means more time to forget.

- You will lose one rotation from your third- and fourth-year clerkships, which will mean less elective time, and perhaps less time to study for USMLE Step 2. You may not be able to take as many away/audition electives or try out specialties that are not offered at your school's affiliated hospital.

- You will largely be isolated from other students as most students move onto the wards by this time. Further, your peers will be celebrating because they are done with the exam, while you are still plugging away. Sometimes, this prompts "extended" students to want to "get the exam over with." This is the exact opposite sentiment you should have if you've extended. You should be diving deep into your weaknesses, spending adequate time to truly understand complex concepts you were not able to grasp during your Intensive Study Period.

The Pros:

- Giving yourself enough time to dive deep into weaknesses that you had not completely understood during the first and second years. Sometimes students need multiple episodes of review before a concept sticks with them.

- Getting a good grasp of clinical topics prior to heading onto the clinical wards.

- Taking the USMLE Step 1 Exam when you are unprepared is a tragedy. Failing the exam and gross underperformance are mistakes that you will never be able to erase. Evaluate why you are in the position you are in. Take responsibility for your actions and do not blame your school, your teachers, your family, or your friends. Once you are able to do this, you will be able to put together the pieces of the puzzle of success.

If you delay your exam but do not fundamentally change what you are doing to prepare for the exam, the benefits will balance out the costs, and your score may not improve. Understand it's normal to be anxious and feel underprepared going into a major exam like the USMLE Step 1. Hopefully you have been using the *Step 1 Method* and are well prepared and know exactly where you are in your prep. Given the opportunity, most students will always opt for more time. But as we've just shown, more time is not always better. You must make an honest assessment of your options and recent performance when choosing to delay your exam.

# Dealing with Failure

Every year, in the late summer months, I receive emails from students I've never met. The subject line of the email usually reads "Help Me: I failed the Step 1 Exam." These students have heard about the *Step 1 Method* at a conference or from friends. Even with our medical school clients, there are a small handful of students who, for one reason or another, are not able to fully implement the *Step 1 Method* and will subsequently fail the exam. Over the past six years, I've coached over 150 students who have failed the exam. Thankfully, with the exception of two or three of the students, all have been successful and have reached their goals on the USMLE Step 1 Exam. They've used the experience as a valuable learning lesson and have gone on to amazing medical careers. This experience has taught me much of the psychology and factors surrounding someone who does not pass the USMLE Step 1 Exam. I am confident that all students who are able to confront and correct these issues can be successful. If you've been able to pass the first two years of medical school, there is no reason you should not be able to succeed on the USMLE Step 1 Exam—with the right training and practice, of course.

I believe that students fail the USMLE Step 1 Exam because of three major factors. Each individual student may have a problem in one or all three areas. If you've failed the exam, you must confront these issues and ask if they are at play.

1.   Deficient Fund of Knowledge

Put simply, the student does not know enough information. This can occur for a number of reasons. The student may not have given themselves enough time to study. They may have been using ineffective methods of study; prime examples are students who spend a majority of their time watching lectures and reading texts. The student may be focusing on trying to memorize as many facts as possible without thinking about the deeper frameworks and connections between topics.

A fundamental problem that occurs with these students is a lack of introspection and assessment. These students tend not to know where they are strong or weak. Because of this, they do not proactively attack their weaknesses. They are overwhelmed with how much information they have to cover and simply do a single pass over all of the information without returning to dive deeper into weaknesses and complicated concepts.

The solution to these problems is to slow down and implement the *Step 1 Method*. Your goal is not to rush through as many questions as possible. You should try to understand concepts at a deep level and be able to anticipate all of the possible questions the NBME can ask you about a particular topic.

2.    Poor Question Strategy

Some students fail the exam despite a good fund of knowledge and strong work ethic. These students tend to have poor a Question Strategy. This category of students collectively term themselves "bad test-takers." The fundamental problem is that the student cannot accurately interpret what the NBME is asking in its questions. These students will unfortunately read the question, then answer a different question in their mind. This universally leads to the student getting the question wrong. They know a significant amount of information about most topics but cannot appropriately access the right information at the right time.

Another problem these students have is the inability to guess effectively. As covered in our Question Strategy section, you should expect to make educated guesses on at least 20–30 percent of the questions on the USMLE Step 1 Exam. Students with question strategy problems tend to shut down when they realize they don't know the answers to questions. They fail to access their adequate fund of knowledge to cancel out answer choices and deduce a best guess. Many of these students will not finish blocks of questions on time. Leaving four to five questions unanswered on a few blocks will force a student to sacrifice 10–20 points that they may have easily obtained.

The solution to all these problems is to develop and practice a question strategy. Our Question Strategy Module and section covers these issues in detail. Many of these students do not practice introspection—they do not ask themselves why they are making mistakes. They will get a poor score on an NBME practice test, but they will not review the expanded feedback provided. It is imperative that students with question strategy difficulties figure out why they got questions wrong on these practice tests. If it is because they simply didn't have the fund of knowledge to answer the question, the solution is simple: learn more about the topic. If it is because they misread the question, they must study the question and learn the pattern. What made them misunderstand the question? Usually it has to do with long questions containing unnecessary detail and distractors.

Finally, the ultimate goal for students with question strategy problems is comfort and familiarity with the USMLE style of questions. You should be able to anticipate in which direction the NBME will go with their questions after you've learned the appropriate frameworks for each topic. Every time you are working questions, you should not be trying just to learn info about the topic, you should be asking yourself: Did I imagine they could ask me this question? Is this is a tricky question? Should I watch out for this in the future? The USMLE Step 1 Exam will have a number of unpredictable hypothetical and experiment-based questions that make up 20–30 percent of the test. Thus, your goal is to gain significant familiarity with the majority of questions which are predictable through question bank completion through the *Step 1 Method*.

### 3.   Lack of Motivation

This last category of factors that contribute to not passing the USMLE Step 1 Exam is the most frustrating. Many students who fall into this category are intelligent and have good test-taking skills but subconsciously are not willing to commit themselves completely to doing well on the exam. A lack of motivation and work ethic can be due to a number of reasons that are complex, varied, and personal. The final result, though, is the same. Some of these students are used to being successful after investing only a minimal amount of effort into studying. They may have done well enough on the MCAT to get into medical school. They may have passed a majority of their medical school coursework. The USMLE Step 1 Exam, though, is a unique exam that must be specifically prepared for. If students do not take the exam seriously, they may find that their minimal preparation will fall short of passing.

Most students see an initial failure on the exam as a call to action. They will assess what worked and what didn't and then reapply themselves to succeeding on the exam. Curiously, there is a subgroup of students that will resent the initial failure on the exam. They will blame their school, advisors, and the resources they used for their failure. They will enter their second Intensive Study Period with a negative attitude and never confront the real issues previously listed. These students are doomed to repeat their performance on a retake of the exam.

For students who find themselves with a lack of motivation, they should ask themselves: why am I in medical school? Many students resent the fact that they must take and pass the USMLE Step 1 Exam. They believe that they didn't come to medical school to take the USMLE Step 1 Exam, but rather they came to learn how to take care of patients and become a doctor. It is true that you must pass the USMLE exams to graduate from med school and become a doctor, but requirements are rarely significant motivators. You should look at the USMLE exams as opportunities to assess your fund of knowledge prior to taking care of patients. No medical student claims that they just want to be an "okay" doctor. The same should apply for the USMLE exam. You should want to do your best and learn as much as you possibly can prior to being responsible for the lives and well-being of thousands of patients. If this is not a significant motivator for you, perhaps you are pursuing the wrong profession.

In my experience, students who have the most difficulty passing the USMLE Step 1 Exam have several conflicting emotions regarding their motivations for wanting to become a physician. This uncertainty definitely translates into difficulty in preparing effectively for the USMLE Step 1 Exam. Confront these issues early on so that you can focus your energies in a highly effective, time-efficient manner.

# Conclusion

Preparing for the USMLE Step 1 exam is a multifaceted experience. There are several aspects of the exam experience that you must anticipate and prepare for. In this text, we've tried to introduce these various concepts and provide strategies for how to excel in them. As we've stated, every student has their own unique potential for success on this exam. This potential will be determined by the time invested to this point, your goals and work ethic, and your openness to trying new methods of studying. We know that most students work hard when studying for the USMLE Step 1 Exam. Our overarching goal is to make sure that every minute you spend studying for this exam is spent in the most high-yield way possible.

Some students take our methods and adjust them slightly to meet their studying style and tendencies. It is okay to do this, but make sure that you never ignore the core principles of the program. Always strive for a deeper level of understanding of topics; never be satisfied with sheer memorization. Understand how the NBME ask questions; be prepared to extrapolate knowledge and think on your feet. Take inventory of your strengths and weaknesses; attack your weaknesses relentlessly and persistently. Understand that doing questions is not enough; you must do questions in a way that leads to a higher level of retention and mastery of the material.

Remember that preparing for the USMLE Step 1 Exam is a marathon, not a sprint. Start early, plan effectively, and live a balanced life. Don't burn yourself out or become frustrated if you stall. If you apply the aspects of the program and are honest with yourself, you will reach your potential, whatever that is. Be patient with the process. Understand that you are not learning this information simply for a score on a test. This information will be crucial to treating your future patients. When you find your motivation waning, take a break and visualize yourself in your future practice. All the hard work you are investing now will make you the best physician that you can be. The choice is yours: will you work to reach your potential on the USMLE Step 1 exam and in your medical career? If the answer is yes, we look forward to helping you reach your goals.

Best wishes and happy studying for the USMLE Step 1 Exam. We know that you can do it!